The Green
Revolution revisited

The Green Revolution revisited

Critique and alternatives

Edited by

BERNHARD GLAESER

Wissenschaftszentrum Berlin
für Sozialforschung
Environmental Policy Research Unit

London .
ALLEN & UNWIN
BOSTON   SYDNEY   WELLINGTON

Allen & Unwin, the academic imprint of
Unwin Hyman Ltd
PO Box 18, Park Lane, Hemel Hempstead, Herts HP2 4TE, UK
40 Museum Street, London WC1A 1LU, UK
37/39 Queen Elizabeth Street, London SE1 2QB

Allen & Unwin, Inc.,
8 Winchester Place, Winchester, Mass. 01890, USA

Allen & Unwin (Australia) Ltd,
8 Napier Street, North Sydney, NSW 2060, Australia

Allen & Unwin (New Zealand) Ltd in association with the
Port Nicholson Press Ltd,
Private Bag, Wellington, New Zealand

First published in 1987

British Library Cataloguing in Publication Data

The Green Revolution revisited: critique and alternatives.
1. Agricultural ecology – Developing countries
I. Glaeser, Bernhard
338.1′09172′4   S482
ISBN 0-04-630014-7

Library of Congress Cataloging in Publication Data

The Green Revolution revisited.
Bibliography: p.
Includes index.
1. Agriculture – Economic aspects – Developing countries.
2. Green Revolution. 3. Green Revolution – Developing countries.
4. Agriculture and state.
I. Glaeser, Bernhard.
HD1417.G723   1986   338.1′09172′4   86-17302
ISBN 0-04-630014-7 (alk. paper)

Set in 10 on 11 point Bembo by Computape (Pickering) Limited
and printed in Great Britain by Billings and Sons Ltd,
London and Worcester.

Preface

In the early 1970s, I conducted research in East Africa on the prospects of introducing ecologically oriented farming (ecofarming) in tropical highlands, paying particular attention to socio-economic conditions and constraints. There, I became acquainted with the Green Revolution and the mighty goals it was attempting. I must confess that I was a little sceptical about this approach, despite the massive boosts in production that were apparently achieved in some areas. In fact, it was during this period that worldwide criticism of the Green Revolution began to develop, on the grounds that it did not improve the lot of the smallest, poorest farmers and peasants, and that the Green Revolution activities had negative effects on local rural environments.

The Consultative Group on International Agricultural Research (CGIAR) responded to the criticisms of the 1970s by creating new programmes. By the 1980s, then, it seemed necessary to dispense with continued criticisms along the old lines and to revisit and re-evaluate the Green Revolution under a new set of conditions. Dr Edmund K. Oasa consented to undertake the task of conducting an initial evaluation of the Green Revolution during his visit to the International Institute for Environment and Society, Science Center Berlin, in 1982. I presented Oasa's paper to Professor Ignacy Sachs, who suggested that the Green Revolution debate be complemented with some relevant case studies, and that Oasa's criticisms be confronted with some positive examples of alternative policies that have been implemented successfully, particularly low-input, ecologically oriented farming systems. Out of the two opposing policies represented by the Green Revolution and the ecofarming approach, this book has evolved.

It is my hope that the results of this study will be met with critical but unbiased interest. To stigmatize one or the other school of agricultural theory is not the intention of this volume: rather, it is to examine the facts of the Green Revolution and to present views which, it is hoped, can ultimately benefit the people in developing countries and their natural environments.

Bernhard Glaeser
Berlin, July 1986

Acknowledgements

I wish to gratefully acknowledge the invaluable assistance given
to me by Mary Elizabeth Kelley-Bibra in the technical
editing and preparation of this book.

Contents

List of tables

1 Agriculture between the Green Revolution and ecodevelopment: which way to go?

BERNHARD GLAESER

The Green Revolution in the 1970s

In 1970 the American botanist, Norman Borlaug, Director of the Division for Wheat Cultivation at the Centro Internacional de Mejoramiento de Maíz y Trigo (CIMMYT) in Mexico, was awarded the Nobel Peace Prize. He was honoured for having set in motion a worldwide agricultural development, later to be called the 'Green Revolution'. This development was based on the genetic improvement of particularly productive plants. Borlaug's so-called 'miracle wheat' doubled and tripled yields in a short period of time. Similar increases were soon achieved with maize and, at the International Rice Research Institute (IRRI) in the Philippines, with rice.

The success of the newly developed strains appeared limitless. They were introduced in several Asian countries in 1965, and, by 1970, these strains were being cultivated over an area of 10 million hectares. Within three years, Pakistan ceased to be dependent on wheat imports from the United States. Sri Lanka, the Philippines, and a number of African and South American countries achieved record harvests. India, which had just avoided a severe famine in 1967, produced enough grain within five years to support its population. Even after the 1979 drought, grain imports were not necessary. India had become self-sufficient in wheat and rice, tripling its wheat production between 1961 and 1980. Such has been the success story of the Green Revolution as propagated by its proponents in the mass media (see, among others, Baier 1984).

Despite its obvious successes, however, the Green Revolution came under severe criticism during the 1970s for ecological and socio-economic reasons (see, among others, Pearse 1977). The main criticism directed against Green Revolution successes was that high yields could only be obtained under certain optimum conditions: optimal irrigation; intensive use of fertilizers; monoculture (for the rational use of machinery and agricultural equipment); and pest control with chemical pesticides (also requiring monoculture).

Further, critics claimed that an important prerequisite for the Green Revolution mode of production was rich soil. Hybrid plants would otherwise be choked out by weeds which have adapted to the less favourable soils, and they could not survive the struggle against insect pests. Moreover, farmers living in problem regions were frequently too poor to be able to afford expensive irrigation equipment and the inordinate amounts of pesticides required (Egger & Glaeser 1984).

Newly discovered environmental calamities and health hazards added more dark colours to an already gloomy picture (Redclift 1984). Intensive fertilization resulted in nitration, in turn causing eutrophication of freshwater streams and lakes. Excessive amounts of pesticides, applied irresponsibly over large areas, created health hazards for rural inhabitants. Moreover, the energy necessary for the production of nitrogen-based fertilizers, for running agricultural machinery, for fuel, and for operating irrigation facilities was severely limited, placing further constraints on Green Revolution development potential.

In the social context, another criticism of the Green Revolution was that it favoured so-called progressive farmers – that is, large landholders with Western education. Small landowners were not subsidized and frequently gave up their farms because they could no longer compete. A landless, rural proletariat was subsequently created; income distribution shifted in favour of the wealthy, and class conflicts developed. Traditional social structures disintegrated so that the extended family, for example, could no longer serve its function of providing social security and care for elderly family members. The resulting mass migration of landless poor from rural areas to the cities has led to the development of urban slums.

The Green Revolution 'revisited' in the 1980s

The Consultative Group on International Agricultural Research (CGIAR) has responded to many of the shortcomings of the Green Revolution. It established the Farming Systems Research (FSR) programme which was designed to cope with the needs of small farmers in particular. Further, it adopted an integrated system of biological pest control; more effort was placed on genetic research as a result of the higher priority given to the pest resistancy of crops, and the efficiency of fertilizers was to be improved. Finally, a number of other new institutes were founded under the auspices of CGIAR, including the International Crops Research Institute for the Semi-Arid Tropics (ICRISAT), established in Hydrabad, India, to cope with production problems specific to arid and semi-arid zones.

In light of these more recent developments (see also Mooney 1983), it seems appropriate to 'revisit' the Green Revolution after a relatively taciturn period in its history to reconsider and re-evaluate its efforts and achievements since the late 1970s and early 1980s.

In 1982, the International Institute for Environment and Society of the Science Center Berlin, commissioned Edmund K. Oasa with the task of re-examining the Green Revolution from a social-science perspective. The objective here was to evaluate the policies of the CGIAR system as explicitly stated or implicitly embodied in authentic documents of the Consultative Group (CG).

The critique and evaluation by Edmund K. Oasa (Ch. 2) is eye-opening: the author subjects Green Revolution policy statements to careful scrutiny; the core of Green Revolution policy is called into serious question, and not just those minor deficiencies or shortcomings that could otherwise be attributed to faulty implementation of policy.

Oasa acknowledges that the Consultative Group has taken new steps and undertaken some novel approaches in response to socio-ecological criticisms of Green Revolution policy. For example, it has instituted a number of projects including the Farming Systems Research programme, a system of integrated pest management, and a programme of biological nitrogen fixation, among others, in an attempt to countermand rural proletarianization. Nevertheless, Oasa argues, the general misery of the poor tends to increase, and class lines and conflicts tend to sharpen as a result of inherent contradictions in CG policies and the politically neutral stance that the Group has adopted, at least superficially. The CGIAR, he writes, 'must take a position of neutrality precisely because it is not neutral'. This implies that

(a) agricultural research develops technology but nothing more;
(b) individual states (countries), not the research institution, are responsible for dealing with internal distribution problems and accompanying inequities;
(c) it is impossible to develop technology exclusively for small farmers; large landowners or corporate farmers will invariably benefit as well.

Oasa argues further that through research's 'inaction', that is, neutrality towards fundamental questions, the objective correspondence between programmes of the state and international agricultural research will become apparent: the predominant, extant class structures will be endorsed and preserved. Unless the state chooses to impose an agrarian-centred development strategy based on small-farm technology, and unless it can provide the capital necessary to accommodate employment for the landless, social tensions are bound to rise, and these could eventually evolve into a national class struggle with redistribution of landholdings at its heart.

In his case study on India, Pierre Spitz (Ch. 3) examines some of India's food systems and discovers that the Green Revolution in India was mainly a 'wheat revolution', patterned after the agricultural models of the industrialized countries. Pierre Spitz questions the appropriateness of this model for India. He makes a plea for 'alternative technologies which could ensure a less uneven development between crops, regions and social groups – reducing disparities between them as well as between seasonal income, employment and year-to-year production'. Thus agricultural research and extension should respond to the 'unverbalized needs of the majority and not only to the effective social demand of the few'. This means that special measures should be taken to eliminate seasonality as a negative factor influencing peasant incomes and the (un)employment of the rapidly growing rural class of landless poor. Spitz suggests that efforts should be made to guarantee cultivation the year round, for example, through dry farming, relay-cropping, or intercropping. This would require some form of institutional support to enable local peasants to organize, not only for confronting technological problems, but also for dealing with issues over land tenure and credit.

Alternative policies

Criticism of the Green Revolution and, in particular, CGIAR's policies – no matter how powerful and convincing the arguments – nevertheless represents only a partial solution directed towards improving the lot of the rural poor in developing countries. The question inevitably arises: is Green Revolution strategy as it is elaborated by the Consultative Group the only direction the alleviation of rural poverty can take? Are there not alternative strategies which have the same end but which employ other means to it?

In the search for alternative strategies, several experts were summoned and challenged to respond to the critiques of Edmund Oasa and Pierre Spitz from their own work and experiences in developing countries on three continents: South America, Asia and Africa. As contributing authors to this volume, each of these experts is in some way or another committed to the concept of an 'ecodevelopment' strategy as a viable alternative. An ecodevelopment strategy is one that (a) is oriented towards fulfilling the basic needs of the poor; (b) promotes self-reliance in agriculture; and (c) strives for environmental compatibility in production methods (see Glaeser 1984).

Ademar Ribeiro Romeiro (Ch. 4) examines development problems in Brazil, where he claims that adherence to the Green Revolution's principles of high-input and labour-saving technology has further aggravated the already existing social inequalities and that new disruptions to the ecology have occurred. The net result of this has been food scarcity and severe rural unemployment in a country whose agricultural area of 323 million hectares (1975) represents 2.5 times the area available for agriculture in India and three times that in China, but whose population is equivalent to only about one-fifth that of India and one-tenth that of China. The author attempts to account for why the country with the biggest agricultural area in the world is unable to feed its own people, yet is extremely dynamic in developing export crops.

The extremely low wages paid to Brazilian workers, caused, in part, by the weakness of the labour unions, is the main cause of insufficient food production. Agricultural policy has played an important role in weakening the workers' capacity to fight for better wages by forcing so many rural inhabitants out of the fields, away from the countryside, and into the cities. Poor living conditions for the urban masses are further aggravated by excessively high prices for food, driven up by a class of extremely powerful and well-organized speculators.

The inability of agriculture to absorb labour results mainly from speculating with the available land rather than using it for production. This is clearly reflected in the high concentration of property in the hands of the few and the waste of much of the available farmland. Most of the land available for agriculture is used as extensive grazing land for cattle. This is the traditional method used to control large areas of land with very little labour input. Often enough, available land is simply abandoned or left totally unused while the landholders wait for prices to rise.

As criteria for selecting an alternative technology, Romeiro proposes that (a) the technology should be compatible with available labour, (b) that it should be energy saving, and (c) that it shall encourage or promote ecological sustainability. This does not imply the return to traditional agricultural practices as such, but it does imply a return to the rationale behind them. The individual production function must be changed in such a way that the production factor of 'capital' can be substituted for by labour and ecological know-how. An abundance of land must be provided for agriculture. Modifications in the production structure must be supplemented with additional economic measures; the purchasing power of peasants and the mass of urban poor must be increased in real terms, and the profits of middlemen and land speculators must be restricted. In short, Brazil's agricultural policy must ensure that ecologically sustainable production patterns are adopted and that the bulk of the 'excess' labour force is absorbed.

In his contribution, R. N. Roy (Ch. 5) claims that 'India's water scarcity and soil erosion problems stem from intensive cropping using unsuitable imported agricultural technology, and from the country's socio-politico-economic reality which has mortgaged long-term sustainability and development for short-term profit involving exploitation of the masses who work on the land'. As an alternative solution for both the socio-economic and the ecological crises in rural India, Roy proposes an ecologically based agro-forestry system which is ideally suited to the fragile soils in arid–humid tropical climates. The author suggests that a farming system analogous to the tropical forest can be used to reclaim eroded land and restore a disrupted water cycle. This would be an organic farming system that protects the soil and recharges aquifers – tree-based and integrated. This concept seems to be economically viable, provided that long-term investments are made available to ensure the survival of local inhabitants during the long gestation periods of forest-based farming. Roy discusses some of the sociopolitical impediments to implementation that pertain to cultural changes and the rural power structure. He proposes pathways to action which include the leasing of waste lands, land reforms, and the mobilization of marginal farmers and landless labour.

The supply and use of energy, particularly in food production, has become a crucial issue in both industrialized and developing countries. Bernhard Glaeser and Kevin Phillips-Howard (Ch. 6) compare the methods of industrialized agriculture with those of an energy-saving technological alternative. Their case study from south-east Nigeria illustrates one such alternative. It describes energy use associated with mixed cropping, looking especially at productivity and the crop-protection aspects of agricultural technology. Glaeser and Phillips-Howard suggest that the ecological knowledge applied in mixed cropping and the energy savings inherent in such a system should be more strongly incorporated in attempts to 'upgrade' traditional agricultural systems and should be given greater consideration in the development of new agricultural policies.

Sustainable agriculture and yet sufficient – shall we say good, average – yields: is this Utopia? Ecodevelopment is an integrated approach which includes the use of cultivation techniques based on the careful treatment and regeneration of resources to intensify and stabilize agricultural production. Such ecofarming systems are appropriate for developing natural production potentials in a sustainable way. Kurt Egger and Brigitta Martens (Ch. 7) describe a project being carried out in Rwanda (East Africa), which puts elements of the ecofarming concept into practice with the primary aims of protecting the soil, improving soil fertility, and integrating animal husbandry into the whole farming system. Ecofarming attempts to combine ecological soundness, production needs, and consideration of the agricultural landscape as human living space into an integrated agricultural system. The overall strategy is to develop and establish sophisticated recycling systems based on high levels of biomass production, achieved through structural and species diversity, ultimately designed to ensure a measure of self-sufficiency. Kurt Egger's experiences in East Africa demonstrate that successfully applied ecofarming techniques are a reality and not merely utopian fantasy, for tropical regions as well as for the temperate zones.

Egger's and Martens' findings are supported by argumentation in Chapter 8 by Bernhard Glaeser, which points out the importance of socio-economic factors as necessary conditions for, or cornerstones of, any technological system designed to enhance agricultural development. Using peasants' crop and farm budgets as a basis, Glaeser analyses the social and economic feasibility of applying ecological production patterns in tropical African highlands and investigates some ways of increasing farm yields. Conclusions drawn in this investigation rest on the premise that the desire for increased crop yields cannot be considered independently from a peasant's basic needs, and that a peasant will incorporate elements of the ecofarming concept into his production strategy once he is convinced of the acceptability of this new technology. This means that a concerted programme of learning, co-operation, in-field training, and mutual exchange of knowledge and ideas between technicians and local peasant farmers will be essential for the successful introduction of new technology, including ecofarming. Bernhard Glaeser's chapter reports on some steps taken towards developing this kind of co-operative exchange, in particular, on establishing pilot training-fields, on planting and maintaining a tree nursery, and on an ecofarming school programme.

Conclusions

A critical view taken of the initial phases and so-called triumphs of the Green Revolution should in no way negate the tremendous potential of agricultural research in general. Rather, this critique should generate a radical reappraisal of Green Revolution objectives and a re-evaluation of the criteria for success.

Ignacy Sachs (Ch. 9) summarizes the criticisms directed at the initial phases of the Green Revolution and proposes some new directions that its international agricultural research could take. One goal, he suggests, would be to obtain production functions which clearly reflect the food–energy nexus: 'Food production ought to be less energy-intensive and, at the same time, more energy efficient.' Another related goal is ecological sustainability. Moreover, while both of these goals have already been considered in ecodevelopment theory and ecofarming research (cf. Chs. 6 and 7 below), the science of aquaculture is still awaiting its own 'neolithic revolution'. The concept of a so-called 'Blue' or 'Aquamarine Revolution' deserves maximum consideration and research priority, on the 'supply side'.

On the 'demand side', Sachs argues, the trend towards developing global uniformity in food consumption habits – 'wheat, bread and noodles, without forgetting coca-cola' – should be reversed. Culture-specific food consumption patterns and ecosystem-specific products and production methods should provide an incentive for combining local peasant skills and rationale with modern, scientific know-how.

Ecofarming is on its way to doing this; but as new developments in the research aims of the CGIAR have shown, so is the Green Revolution, at least as far as technical programmes are concerned. Unfortunately, however, the social compatibility of new technologies tends to be a neglected area of concern. This has definitely been the case with the Green Revolution (cf. Ch. 2 below), but the danger is also inherent in the practice of ecofarming, since the problem is primarily not the inadequacy of food supply, but the inability to purchase it.

Policy recommendations must, then, include measures for dealing with major social problems (in particular, distributional problems) and ecological shortcomings (environmental problems). Socially responsible research must not cling to the pretence of political innocence and shy away from the basic (and particularly unpleasant) facts: research policy, Green Revolution or ecodevelopment-oriented, must acknowledge that it has a service function to fulfil for individual consumers and for producers. Food production must be labour-absorbing and environmentally sustainable, rather than capital-intensive and energy-wasting. Or, as Ignacy Sachs so aptly puts it, 'More people must be able to earn their access to food while working to produce it.'

References

Baier, W. 1984. Die 'Grüne Revolution' hat die Ernährungslage deutlich verbessert. *Frankfurter Rundschau*, 2 January.

Egger, K. and B. Glaeser 1984. Kritik der Grünen Revolution: Weg zur ökologischen Alternative. In *Ökologischer Landbau in den Tropen*, P. Rottach (ed.), 13–37. Karlsruhe: C. F. Müller. 2nd edition 1986.

Glaeser, B. (ed.) 1984. *Ecodevelopment – concepts, projects, strategies*. Oxford: Pergamon.

Mooney, P. R. 1983. From Green Revolution to gene revolution. In *The law of the seed – another development and plant genetic resources. Development Dialogue* **1–2**, 84–94.

Pearse, A. 1977. Technology and peasant production: reflections on a global study. *Development and Change* **8**.

Redclift, M. 1984. *Development and the environmental crisis: Red or Green alternatives?* London and New York: Methuen.

PART I

The Green Revolution: policy and implementation

2 The political economy of international agricultural research: a review of the CGIAR's response to criticisms of the 'Green Revolution'

EDMUND K. OASA

Introduction

This chapter is about an integrated circuit of 13 research organizations which develop and/or support new agricultural technology for the so-called Third World. Commonly referred to as international agricultural research centres, they are funded by an informal club of donors known as the Consultative Group on International Agricultural Research (CGIAR or CG or the Group), which is co-sponsored by the World Bank, the United Nations Development Programme, and the United Nations Food and Agriculture Organization (see Table 2.1 for list of donors). This research system's accomplishments have not gone unnoticed as they were the target of many critics in the 1970s. Critics argued that the introduction of new high-yielding, short-statured, nitrogen-responsive cereals developed by two of the oldest centres, the International Rice Research Institute (IRRI) and the International Centre for the Improvement of Maize and Wheat (CIMMYT), widened inequalities and hastened rural landlessness and environmental decay. In recent years, international agricultural research has responded. Our task is to examine this response and its meaning for world hunger and poverty.[1]

Since the heralding of the 'Green Revolution' (the expression that popularized the work of IRRI and CIMMYT), the start of several more international centres located in other parts of the Third World

Table 2.1 Membership of the Consultative Group on International Agricultural Research (January 1983).

Countries	International organizations	Foundations	Fixed-term members representing developing countries
Australia	African Development Bank	Ford Foundation	Asian region: Indonesia and Pakistan
Belgium	Arab Fund for Economic and Social Development	International Development Research Centre	African region: Senegal and Tanzania
Brazil	Asian Development Bank	Kellog Foundation	Latin American region: Colombia and Cuba
Canada	Commission of the European Communities	Leverhulme Trust	Southern and Eastern European region: Greece and Romania
Denmark	Food and Agriculture Organization of the United Nations	Rockefeller Foundation	Near Eastern region: Iraq and Libya
France	Inter-American Development Bank		
Germany	International Bank for Reconstruction and Development		
India	OPEC Fund		
Ireland	United Nations Development Programme		
Italy	United Nations Environment Programme		
Japan			
Mexico			
Netherlands			
Nigeria			
Norway			
Philippines			
Saudi Arabia			
Spain			
Sweden			
Switzerland			
United Kingdom			
United States			

Source: Consultative Group on International Agricultural Research, Secretariat, 1818H. Street NW, Washington DC 20433, USA.

has represented a rapidly evolving institutional framework for the internationalization of agricultural research. The status of Rockefeller Foundation country programmes in Latin America, which began in 1941, changed to international operations conducting multi-disciplinary research. By 1969, a loose federation of four centres existed: in addition to IRRI and CIMMYT, the Rockefeller Foundation in co-operation with the Ford Foundation had established the Centro Internacional de Agricultura Tropical (CIAT) in Colombia and the International Institute for Tropical Agriculture (IITA) in Nigeria. In 1971, the Consultative Group was formed to continue support for these four centres and to consider the start of additional organizations.

Since 1971, the CGIAR has added nine organizations to its circuit. According to CG documents, most of them were largely meant to fill the gaps that research on rice and wheat could not cover. For example, in 1971 the Group established the International Crops Research Institute for the Semi-Arid Tropics (ICRISAT) to do research on commodity and farming systems for a particular ecological zone of the Third World. It also filled the void in livestock research in 1973 by approving the construction of the International Laboratory for Research on Animal Diseases (ILRAD) and the International Livestock Centre for Africa (ILCA).

However, the institutionalization and expansion of international agricultural research was occurring at the same time as criticisms of the Green Revolution surfaced. These criticisms will be discussed in more detail below, but the following questions arise here: were these institutes and respective programmes set up solely to compensate for technological and geographical shortcomings? Is there a connection between their formation and the criticisms? And with these questions in mind, what have the main actors in this once-hailed strategy done to respond to the criticisms?

The application of modern science to Third World agriculture is one part of an effort to eliminate hunger and poverty. There is, of course, a range of projects and programmes falling under the heading of rural development, such as projects for infrastructure, small-farm credit, and irrigation, to name a few. While these form bits and pieces of a larger picture, agricultural research has grown in stature over the years in this maze of projects. This is best shown by the 1981 World Bank sector policy paper, in which the Bank indicated a rise of $200 million between 1979 and 1984 for agricultural research and extension. Underlying this increase is the fact that 'the proportion of Bank resources allocated to agricultural research is likely to grow more rapidly than that committed to extension'.[2]

A major rationale for this increase is the ultimate relationship between research and socio-economic development. The second review carried out by CGIAR in 1981 found that 'there is a greater recognition that *development* is the ultimate purpose of supporting international agricultural research' (emphasis added). The Bank reaffirmed this position by stating that '*how benefits are distributed among farmers and the extent to which they are shared by others depnd largely on the characteristics of a country's research effort*' (emphasis added).[3]

During moments of crisis, social phenomena (production systems, organizations, institutions, individuals) reveal their 'stuff', so to speak, in terms of what is possible or what is to be done. For international agricultural research, the criticisms and reported consequences of the Green Revolution brought to an end a brief period of euphoria and a 'crisis of understanding'. This chapter analyses the following responses of research.

First, international agricultural research acknowledged and incorporated into its research agenda certain arguments presented by the Green Revolution critics. One more recent observer, Pat Roy Mooney, revealed a statement by Henry Romney, Director of Information for the Rockefeller Foundation, who said that 'much of the criticisms of the past few years have been accepted. The critics have turned us around.'[4] The second deals with the precise *content* of this about turn. In brief, the acceptance of the criticisms and the about turn resulted in a refined research agenda based on the concepts of distribution and equity. I will argue that the employment of these concepts and associated research projects detract our attention from the fundamental problematic that critics pointed to: the social organization of capitalist agriculture.

Third, owing to the partial nature of the about turn the response of international agricultural research is internally plagued by contradictions and tensions that certain sections of this integrated circuit seem to recognize. Finally, I will argue that these contradictions are unresolvable owing to research's rigid position towards the consequences of the new technology. To establish this argument, I will refer to the relationship of research to capitalist state apparatuses in the Third World and its consequent location in the widening social polarities, to which criticisms were directed. In short, it is not happenstance that research's assessment is incomplete. Instead it is a politically bound assessment and one that corresponds to a 'neutral' relationship with the state.

To these ends, I will draw freely and primarily from the documents of the CGIAR, statements of its donors, and reports and reviews of and by its centres. These will speak for themselves.

An introductory note on the Consultative Group on International Agricultural Research

Although formalized under the Consultative Group on International Agriculture Research in 1971, international agricultural research already had a considerable history. As stated earlier, four centres were operating prior to 1971: IRRI, CIMMYT, IITA and CIAT. The accomplishments in rice and wheat, in particular, justified expanded support for the four existing centres. In the longer view, however, this history also included the vision of the Rockefeller Foundation, which started a series of events beginning in 1941 in Mexico that culminated in the formation of the Consultative Group. As early as 1954, J. George Harrar, a major figure in formulating plans for Foundation activities in Mexico and the Philippines, articulated a vision of internationalizing agricultural research during an address to the North American Wildlife Congress.[5] For Harrar and the Foundation in general research was destined to become the leading factor in eradicating world hunger. The agricultural sciences knew no national boundaries.

According to the CGIAR's *Report of the Review Committee* circulated in January 1977, the Group had 16 donors who contributed $20.06 million in 1972. By 1976, the number of donors rose to 26 and the budget to $64.39 million. For 1981, its Integrative Report for 1982 indicated a total of 40 donors and an estimated 'core' budget of $157.945 million.[6]

While these figures show consistent increases, the proportion of the centres' budget of total multilateral and bilateral assistance for agricultural research devoted to the Third World has remained fairly constant: 5.2% in 1970, 8.1% in 1975, and 8.7% in 1980.[7] These figures may also appear surprisingly low when determining the importance of the CGIAR as a general institution of international agricultural research. However, while considering the Consultative Group's geographical–ecological coverage, the range of commodities its centres attempt to improve biologically, their network of contacts with research and governmental organizations throughout the world, and finally, the centre's stature in prominent assistance agencies, the Consultative Group and the international centres remain at the core of widespread activity.

The period 1971–6 was one of geographical as well as financial expansion. In 1972, the Group provided funds for the start of the International Potato Centre (CIP) in Peru, which was an outgrowth of a previous Rockefeller Foundation programme begun in the 1950s. Also in that year, a memorandum of understanding was signed with the Indian government to start the International Crops

Research Institute for the Semi-Arid Tropics (ICRISAT) in Hyder-abad. In 1973, after considerable discussion, the Group decided on monies for the International Laboratory for Research on Animal Diseases (ILRAD) in Nairobi, Kenya and the International Live-stock Centre for Africa (ILCA) in Ethiopia. And in 1974, the International Board for Plant Genetic Resources (IBPGR) was founded and situated in Rome, Italy. Funding in that year also started for an activity of a different character but deemed worthy of attention – the West African Rice Development Association (WARDA). With headquarters in Monrovia, Liberia, WARDA is a co-operative programme covering 15 West African countries dealing with rice research and development. Finally, in 1976, the International Centre for Agricultural Research in the Dry Areas (ICARDA), located in Beirut, Lebanon, was approved for inclusion into the CGIAR system.

Because the system expanded rapidly during this period, the Group agreed to a review of the system, which declared a three-year period of consolidation for funding and administrative procedures as well as allowing the newly established centres to mature. The committee, comprised mainly of representatives from major donor agencies and directors of international centres within the CGIAR system, reaffirmed the system's 'focus on research and technology development related to food commodities widely consumed in the developing world';[8] recommended that 'continued support should be provided for the current set of centres and related activities';[9] and cautioned the CGIAR 'about assuming responsibility for major new activities such as direct support for extension efforts or national programs'.[10]

Although in a period of consolidation, the system did undergo financial growth of about 16% in real terms per annum. The post-consolidation period brought about a modest expansion, with the creation of the International Service for National Agricultural Research (ISNAR) and the inclusion of the already formed Inter-national Food Policy Research Institute (IFPRI), which brought the system to its 1983 status.

In 1981, the CGIAR underwent a second review. From its vantage point, the system had progressed to a point where it had to consider issues of cost effectiveness; the system's place and status within the larger international development assistance picture; and new areas of research given the perceived strengthening of national research programmes. Again composed of members 'chosen for their personal experience of the CGIAR system and its com-ponents',[11] the review committee reaffirmed the system's focus on specific commodities and cropping systems research based on them;

encouraged the centres to engage in more forward planning; provided procedures to streamline administrative activities; and recommended that the international centres reassess priority areas within the context of stronger national programmes. 'It may be possible', it suggested, 'to phase out work on some of the existing commodites, thus releasing resources for other work.'[12]

While at least three periods are discernible in this history, it is important to note a stated change in 'research philosophy'. It is a change that I believe reflects more than the Group has indicated. Both review reports recognized the 'major breakthrough mentality' characterizing our subject's early history. This mentality was based, the second review report stated, on the ideas of J. George Harrar and Forrest F. Hill of the Ford Foundation, both of whom engineered the formation of the International Rice Research Institute in the late 1950s. According to the same review, the 'Hill–Harrer philosophy' went as follows:

> The essence of the Hill–Harrar philosophy was that the Centres should be independent, autonomous and highly efficient research institutes, generating urgently needed technology that would take too long to generate through national research programmes. *The key to this philosophy was the time factor. It was recognized that research capabilities in the developing countries would grow and would increasingly contribute to the total research effort but, faced with the reality of the world food problem, it was essential to take immediate action, and to concentrate the effort in order to obtain quick results.*[13] (Emphasis added.)

The first advances that resulted in the expression 'Green Revolution' were considered 'quantum jumps' based on the simple, straightforward goal of increasing yields, especially in ecologically favourable areas having controlled irrigation.[14] Contrary to the 'major breakthrough mentality', the system has expected, at least since its rapid expansion in the first half of the 1970s, that future science-based technological advances for agriculture would be cumulative and incremental.

The reason for this apparent switch is that international agricultural research has turned its attention to unstable environments, in which soil is less fertile and more acidic, land is dependent upon rainfall instead of irrigation, and where the 'small, resource-poor farmer' lives.[15] Research cannot be expected, the consensus seems to be, to yield quick, visible results from conditions which are far from predictable and optimal. Research, in this situation, has to get down to the cultivator's world, understand the environment, and then propose agro-technical changes tailored to that environment, including what each farmer can or cannot afford.

However, the link between unstable environments and the resource-poor farmer tells us a great deal more about this switch in philosophy. It tells us that the switch is not solely technical, biological, or environmental. For throughout the three periods of CGIAR history, there was considerable and frequent discussion about the plight of small farmers, both landed and landless, and the need to improve their farming systems as a second stage of the Green Revolution. In conclusion, the connection between the small farmer and the marginal, suboptimal environments has determined how this research system has responded to criticisms of its work.

Criticisms of the 'Green Revolution' and the political economy of new agricultural technology

Before proceeding to our central discussion on the Consultative Group and its centres, it will be helpful to recall the essence of the criticisms of this food strategy. The most comprehensive work on this topic is a set of country-specific and general studies done by individuals who were connected to the United Nations Research Institute on Social Development (UNRISD).[16] We can refer to these reports, which have been summarized by the project's leader, the late Andrew Pearse, and a couple of others as well.

According to Pearse's 'Technology and peasant production: reflections on a global study', the UNRISD investigators were required to pose questions of power and its organization in terms of class relations. While each devised his/her own methods of research and measurement standards, 'field studies were required which would approach the livelihood and social relations of the different groups and classes of persons involved in agricultural production' as well as government programmes, processing and marketing systems, and relevant agrarian institutions. Pearse described the project's main conclusion in the following terms:

> [T]he message of these reports and the conclusion drawn from the whole experience is that the technology will inevitably play an increasingly important part in agricultural production, *but the main principles of the strategy adopted for introducing the technology are inadequate for the development needs of the mainly rural countries concerned and harbour a potential for increased pauperization and social conflict.*[17] (Emphasis added.)

These principles, Pearse said, referred to the requirements imposed upon peasant cultivators to be agronomically dependent upon seeds and inputs developed by forces distant from the 'self-sustaining local production/consumption systems'. The particular technological forms also required cultivators to seek loans in order to adopt them profitably. These two requirements meant that the cultivator must 'reorganize his economy' to produce a minimum surplus to pay for the biological inputs and the interest due on the initial loan. Consequently, Pearse reported, this reorganization process, in which farmers had to become businessmen skilled in market operations, was too much of a change, resulting in indebtedness and, for some, landlessness.

In this context, the 'terms of incorporation' for the various actors in agricultural production become important. These terms point to the social processes that involve the formation and reproduction of classes and groupings which reflect the distribution of power in agriculture. Citing two of the UNRISD studies, Pearse had the following to say:

> Perhaps the most important aspect of the rural situation ... is the fact that rich and poor do not simply co-exist. The accumulation of land by the rich creates a demand for labour which the poor are obliged to satisfy because of their land-poverty or landlessness; moreover, the entrepreneurial success of the rich is made possible by the hunger and importunacy of the poor cultivator who is obliged to surrender his bargaining freedom and even to pledge his future labour at a reduced price in order to sustain his family and meet current obligations.[18]

While the rise in landless wage labour, which we will call rural proletarianization, would eventually be recognized by certain sections within international agricultural research, more recent statements have reaffirmed the project's findings. An article by T. J. Byres entitled 'The new technology, class formation, and class action in the Indian countryside' is exemplary. Byres reviewed the specific nature of the new technology, de-mystified the separation between biochemical and mechanical technologies, and noted their effects upon rural labour. Examining the relationship between India's agrarian structures, the Indian state, and the new technology in certain sections of the country, the author noted the consolidation of the 'rich peasantry as a powerful dominant class: the rich peasantry has become stronger economically and has taken on more of the characteristics of a class of capitalist farmers'.[19]

As a logical and necessary outcome, the new technology has contributed to proletarianization, albeit partial. By way of example, although figures for the areas of the Punjab and Haryana have shown a decrease in households owning no land at all and a rise in the proportion not cultivating the land, Byres argued that this actually represented partial proletarianization. 'In other words',

Byres said, 'the poor peasant must, increasingly, sell his labour power in order to survive', although still owning some land.[20] A poor peasant in this context is not dispossessed of land in the total sense but may be involved in a situation of 'tenant-switching', wherein a small landowner leases out land to the expanding class of rich peasants and simultaneously works the land as a labourer.[21]

Byres' reported developments reveal that the destiny of an enlarged rural workforce will depend on the generation (or absence) of employment. The crisis of labour is central to the above analyses and, as will become pertinent below, to the response of international agricultural research. The inextricable relationship between technological forces and the determining social context of production relations is clear. To be sure, the precise forms of these consequences, which are tied to the specific character of the production process, cannot be explained solely by the introduction of new technology. Technology, instead, has contributed to and/or hastened a social process set in motion prior to the internationalization of science-based agricultural research.[22]

The arguments outlined in the preceding paragraphs apply to other and more recent areas of criticisms, namely, the environment and the ecology of agriculture. An extensive literature exists on the depletion of the world's strategic resources, on the dangers of pesticides, and on the rapid erosion of topsoil that takes years to replenish.[23]

Regarding the Green Revolution, a first thought that comes to mind is the extinction of valuable germ plasm owing to the introduction of crop monocultures that typified the early stages of international agricultural research. This danger points to another threat found in the genetic vulnerability of modern agriculture. We need only to recall the cornleaf blight of 1970, which left the US southern states with only half a harvest.[24] According to Kenneth Dahlberg's *Beyond the Green Revolution: the ecology and politics of global agricultural development*, the epidemic was caused by a new race of leaf blight fungus which is damaging to varieties possessing the Texas male sterile cytoplasm. The same author reported that in 1970, about 90% of hybrid corn in the US had this particular cytoplasm because 'it saved the cost and inconvenience of hand-detasselling for seed production'.[25] This condition of genetic uniformity applies, for example, to the high-yielding varieties (HYVs) of rice; a report put out by the International Rice Research Institute indicated that a large proportion of high-yielding rice varieties throughout the world contain the same maternal parent, Deo-gee-woo-gen, which is an Indonesian variety.

Related to genetic vulnerability are problems associated with plant resistance to insects and diseases. Insects that were known to be obscure and relatively harmless before the introduction of HYVs have proliferated due to a host of factors. Moreover, certain types of plant resistance break down in a short time to predominant insects, which are able to generate their own biotypes. This complicates the plant breeding processes necessary for resistance development as well as complicating the ecology of insects.

The list of environmental and ecological consequences goes on to include the long-term effects of exogenous fertilizer. 'In many irrigated areas', Dahlberg said, 'applications of fertilizer have reached a point of diminishing returns that it would take 6.5 times as much fertilizer as now applied to double their production.'[26] This alone may be the limiting factor in HYV production unless other practices are developed to increase fertilizer efficiency. Finally, fertilizer-based crop intensification has been associated with water contamination by nitrates and phosphates and oxygen reduction.

Concluding our checklist, there are the many issues related to soil and land use, which include deforestation, soil erosion, the generation of dust through agriculture, and overcompaction from heavy machinery, to name a few. On the surface, these problems as well as those noted above appear to be purely technological and, therefore, begging for appropriate technological solutions. But, as Dahlberg states for us, these observations 'go back to the central position of ... the distribution of power, influence, and economic goods'.[27]

Certainly, efforts are underway to cope with environmental decay and depletion of the world's resources. While scientists and other individuals have deplored the hastening pace of these historical processes, efforts to slow them down or to avoid them by developing different but 'appropriate' technological forms will ultimately neutralize and depoliticize the problem that presently concerns us. The intensified pace of these ecological processes has occurred alongside the more commonly discussed social consequences in the form of rural proletarianization and social differentiation in capitalist agriculture. Fundamental to the discussion of technology and its consequences is the organization of production and its social relations. To speak of one is also to speak of the other. By necessity, this connection points to the research process itself. If increasing social polarity characterizes both context and consequence of technology, then how is agricultural research connected to it? This question will become more prominent in the following sections.

Technology for the 'resource-poor' farmer: an attempt to resist rural proletarianization

International agricultural research entered the 1970s with an awareness of a rising body of literature on its work and on the performance of Green Revolution technology. Research's response was eventually to incorporate the charges of critics into its agenda but, as we stated earlier, on its own (technological) terms. What were these terms of reference? This section will discuss the response of research to increasing rural proletarianization and review associated projects and programmes which first, have been prominently noted in documents prepared by the Consultative Group, the Group's Technical Advisory Committee, and the international centres, and secondly are designed to resist this historic social process. These projects include Farming Systems Research, Integrated Pest Management, and Fertilizer Efficiency.

The older, more established institutes, particularly IRRI and CIMMYT, bore the brunt of the critical literature. Of the two, IRRI took the lead in responding to criticisms with a special project devoted to studying the consequences of new technology.[28] In a short time, IRRI's experience in dealing with these consequences became a starting point for CG attention and treatment.

By 1975–6, the rice institute had expressed interest in the matter on a number of occasions. One occasion was an IRRI-sponsored conference in 1969 on rice research and training for the 1970s, which was partly motivated by the desire 'to examine the economic and social consequences of new production technology'.[29] In 1971, IRRI sponsored a multi-village survey in the Asian region that discussed among other topics the (mal)distribution of benefits and consequent potential for social conflict.[30] Developments like these culminated in a 1976 conference, the purpose of which was to 'put together the empirical findings of IRRI research and to promote positive discussion on this controversial issue using IRRI research findings as background materials'.[31]

These events legitimized the attention of research to the consequences of the new technology. Yujiro Hayami, an IRRI economist, directed at IRRI a series of village-level studies entitled 'Anatomy of rice village economy' and 'Dynamics of agrarian change'. These studies posed the following questions:

> Is the system viable in the long run? If present trends continue, farm size will decline further, landless laborers will continue to increase in numbers relative to farmers. Real wages will decline and the value of tenancy rights will rise widening the income gap between farmers and landless workers.[32]

The growing concern that rural proletarianization had become a 'problem' was apparent. Widening gaps in income coupled with declining wages were couched in the context of an evolving polarity between actual farmers and a landless workforce. The fledgeling relations of production, however, could only be aggravated by declining wages, thereby preparing the ground for political strife.

Economists within the CGIAR system eventually expressed this danger in non-centre publications. In 1978, two noted economists connected to international agricultural research assembled a set of essays, some of which dealt with the earlier criticisms. In one essay, 'Induced innovation and the Green Revolution', Hans P. Binswanger, an economist, and Vernon W. Ruttan, IRRI's first agricultural economist in the 1960s, stated the following:

> It does seem clear, however, that the contribution of the new seed-fertilizer technology to food grain production has weakened the potential for revolutionary change in political and economic institutions in rural areas in many countries in Asia and in other parts of the developing world. In spite of widening income differentials, the gains in productivity growth, in those areas where the new seed-fertilizer technology has been effective, have been sufficiently diffused to preserve the vested interests of most classes in an evolutionary rather than a revolutionary pattern of rural development.
>
> By the mid-1970s, however, the productivity gains that had been achieved during the previous decade were coming more slowly and with greater difficulty in many areas. Perhaps revolutionary changes in rural institutions that the radical critics of the green revolution for the past ten years have been predicting will occur as a result of increasing immiserization in the rural areas of many developing countries during the coming decade.[33]

Attention to rising landlessness escalated in both centre and CGIAR discussions. IRRI sponsored a workshop in 1978 on village-level studies and invited social scientists active in the field to take part. Participants expressed concern over small farmers and labourers losing control of their own futures and not being adequately represented by government research and development programmes. Annual statements (published by the CGIAR Secretariat) of CG discussions in the late 1970s frequently had sections devoted to the problem of distribution and the needs of the small, resource-poor farmer.

The Consultative Group's 1977 Integrative Report discussed how complex research had become for its research system owing to the unstable farming environment and the socio-economic position of the small farmer. It further stated that research 'methodology and time frame are much less understood' when research has as its objective the social development of the small farmer. In stating this, the report presented the views of several donor representatives who felt that:

> the benefits of research should flow more specifically to the poor in the LDCs [less-developed countries], hence the need for priority to research which would benefit the small farmer and rural landless and produce low risk, inexpensive technology that can be applied quickly.[34]

Given this objective, the report expressed difficulty in setting a priority for the allocation of resources. The Group itself (referring specifically to donors) encouraged priorities to be set in terms of 'target groups', while its Technical Advisory Committee favoured setting priorities along commodity lines. The centres, however, seem to have done both, with the additional criterion of research priorities set according to ecological, agro-climatic zones. These differences, however, are not at all crucial. Donors, TAC and the centres generally agree that the system's current research programme fulfils the stated criteria: small farmers are indeed situated in less-endowed and more uncertain areas of production. The important point for us is that the CGIAR's post-1976 consolidation period placed the small, resource-poor farmer at the forefront of agricultural research.

The terms of reference mentioned earlier can be seen in this emphasis on the small farmer, which is an outcome of the assessment by research of the consequences of its work. The assessment defines social conditions and processes (i.e. proletarianization as consequential) in terms of distribution and equity. This assessment is clear in the following statement by the CGIAR:

> [r]egardless of the transitory nature of this regressive impact on relative income distribution, there will always be a permanent widened difference between the levels of total income. As long as there is inequality in the ownership of productive assets and large farmers own more land than small farmers, larger farmers will gain more income ... Therefore, *in the absence of widespread and effective distribution of productive factors, absolute income differences must widen.*[35] (Emphasis added.)

The extension of these arguments ultimately points to our position that (mal)distribution is symptomatic and inextricably tied to the more fundamental problematic of the organization of production and its contradictory social character. The fact that this extension does not occur is not accidental but necessary and significant in historical and political terms – terms to be articulated in a later section. In the CG's position, the concepts of distribution and equity result in a redefined agenda that narrows its technological

mandate to aim at a particular target population. The failure to
extend its own assessment to a logical conclusion (that is, the social
contradictions of production) indicates the position and role of
research *vis-à-vis* proletarianization – namely, the goal of resisting or
constraining it. Research programmes and projects, as shown
below, can only correspond to this position by attempting to make
farming attractive and profitable for the resource-poor farmer.

The same set of CGIAR documents referred to earlier has also
suggested the tenuous character of this re-defined agenda. The 1977
Integrative Report speculated that 'it seems virtually impossible to
design a technology that helps the poor farmer and not the rich'.
This view was again expressed in the following year's report (1978):
'The review of the literature has produced little evidence that
technologies have been successfully designed exclusively for the
small farmer.' 'Even if such technology were developed', the report
continued, 'large producers would very likely adopt whatever
attributes of "smallness" that were needed to apply such technologies if it were to their advantage.'[36]

Despite these reservations about small-farmer research and development, the CG and its centres have insisted on maintaining current
policy. At a 1978 meeting, the centre directors and Group donors
voiced the following concerns:

1   A speaker emphasized the need to stress research on increased
    production benefiting the resource-poor farmers and the rural
    poor. He felt that the problems of such people had not up to
    now been adequately addressed by scientists or development
    planners and felt that they should be the primary target of the
    Group's efforts.
2   Another speaker drew attention to the fact that research to serve
    the resource-poor farmers is likely to need more money and
    more time. He, as representing a major donor, accepted this
    and expected each center to factor these considerations into its
    program decisions.
3   One speaker thought the [1978] Integrative Report demon-
    strated a trend towards equity and away from production. His
    own authorities welcomed this.
4   Another speaker, joining those who emphasized equity con-
    siderations, pointed out that knowledge of the development
    process itself, particularly in rural areas, was still very limited.
5   Another speaker, in welcoming the evident shift towards more
    emphasis on equity, felt that it had not gone nearly far
    enough.[37]

The Consultative Group and its centres have not attempted to resolve their own scepticism. This lack of attempt is necessary and logical given the incomplete assessment of consequences discussed above and the relationship of research to other classes and groupings in the dominant social order of which proletarianization is a distinct part. Our present point is that the attempt by the CG to respond to criticisms of its work – criticisms that have been incorporated into its agenda although in partial form – has been within its own terms of reference expressed through the concepts of equity and distribution. These terms essentially open the door for the corresponding programmes and projects described below.

Understanding the cultivator's world: Farming Systems Research (FSR)

In 1977, a three-person team appointed by CGIAR–TAC reviewed research into farming systems at the Centro Internacional de Agricultura Tropical (CIAT), the International Institute for Tropical Agriculture (IITA), the International Crops Research Institute for the Semi-Arid Tropics (ICRISAT), and the International Rice Research Institute (IRRI). The summoning of this review not only pointed to the technical complexities of FSR but also its centrality to international agricultural research's objective of small-farm development. The introduction to the review document reaffirmed research's focus on the 'many millions of small farmers' and increased productivity in the small-farm sector.

What is FSR? Attempting to be both broad and specific, the review team presented the following definition of a farming system:

> A *farming system* (or farm or whole-farm system) is not simply a collection of crops and animals to which one can apply this input or that and expect immediate results. Rather, it is a complicated interwoven mesh of soils, plants, animals, implements, workers, other inputs, and environmental influences with the strands held and manipulated by a person called the farmer who, given his preferences and aspirations, attempts to produce output from the inputs and technology available to him. It is the farmer's unique understanding of his immediate environment, both natural and socio-economic, that results in his farming system.[38]

Research's relationship to a farm system, then, is first to understand the interrelationship between elements of a system and, second, 'to enhance the efficacy of farming systems through the better focusing of agricultural research so as to facilitate the generation and testing of improved technology'. Central to these steps seems to be the near-preservation of a small cultivator's farming system. '*In this* [FSR] *lies a major challenge for the IARCs*', the team said. '*Unless their research*', it continued, '*leads to applicable technology that is incorporated by the farmer in his farm system, research resources will have been wasted*' (emphasis in original).[39]

The aforementioned review document and others cited below will further reveal the intimate connection between FSR and small-farm development; and secondly recapitulate the tension and contradiction of this connection which we discussed in the previous paragraphs on the CGIAR's general thrust. Without reviewing an entire history of FSR, the connection between FSR and small-farm development was established soon after the critical literature surfaced. The IRRI experience is instructive in that the Institute saw the move into FSR as a logical next step in its programme for the 1970s.[40] When IRRI's administration changed hands in 1973, the new director-general, in co-operation with two other prominent figures in international agricultural research, designed a proposal that highlighted the potential of 'cropping systems research' to 'get down into the cultivator's world' where there was little or no irrigation and sub-optimal growing conditions.[41]

TAC reviews of CIAT, IITA, ICRISAT, and IRRI reported this connection. From 1973 to 1975, CIAT's major programme was the Small Farm Systems Programme, which aimed at understanding the great diversity of farming systems in tropical Latin America and focused on family farms as integrated systems.[42] Although the programme was disbanded in 1975,[43] the small-farm thrust remained, but along commodity lines.

In 1976, TAC reported that CIAT's commodity programmes in cassava, beans, beef, rice and maize were 'respecified' to ensure that production technologies would be suited to *all* farm sizes, 'including those with limited resources'.[44] Although without a formal FSR programme and despite TAC's statement, CIAT's respecified focus is based on a farming systems approach, in that the centre has been attempting to fit new technology into production systems of small farmers. In its long-range plan released in November 1981, CIAT reported that because 'the large farm sector is already well-served with agricultural production technology', its staff decided that it should concentrate on expanding the agricultural frontiers (in other words, previously marginal lands) and on small-farm intensification. Its programme for the 1980s reaffirmed the words of CIAT's director-general, John L. Nickel, who had said earlier that 'improving the welfare of the poor will be a job not of a special team, but of every member of the staff'.[45]

While CIAT does not have a formal FSR programme, IITA and

ICRISAT do. Farming systems research is found first on IITA's agenda, followed by commodity programmes that feed into it. According to the TAC quinquennial review completed in 1978, IITA stated that the small farmer in the humid and sub-humid tropics is the main target of research. Its efforts take the form of 'finding ways and means through national programmes and other collaborating agencies to reach the farmer[;] and of increasing productivity with a low level of cash inputs by developing improved material and cropping patterns'.[46]

The notions of minimizing inputs and being resource-minded are also prevalent at ICRISAT, whose FSR programme is described as 'resource-centred' and 'development-oriented'. One of its objectives reveals the small-farmer thrust: 'to generate economically viable, labor-intensive technology for improving and utilizing, while at the same time conserving, the productive potential of natural resources' (Technical Advisory Committee 1978a, p. 10).

The content of various centres' mandates indeed suggests a switch from earlier ones in international agricultural research history. The straightforwardness of the early IRRI and CIMMYT programmes, which were based on the major breakthrough mentality with little or no reference to environmental and conservation practices, has been declared almost obsolete in recent CGIAR documents and has been replaced by a more incremental approach found in such programmes as FSR. Moreover, FSR has brought about a rise in stature for socio-economic research by integrating the works of agricultural economists and, to some extent, anthropologists into improving farm technology and practices. These social scientists have contributed 'baseline' data and information on 'constraints' on adoption and yield potential, mechanization possibilities, and so on. Fully consistent with this research system's general response to the consequences of the Green Revolution, the rise of socio-economic research represents the attempt to gauge beforehand the probability of failure of new farm technologies that could contribute to farmer indebtedness and dispossession.

On this note, FSR appears to represent a significant development in international agricultural research. Stressing the requirement of understanding the small farmers' environment, FSR with its socio-economic component is an attempt to deal with the charge that farmers have been losing control of their destinies and decision making. It emphasizes a closer relationship between research and the farmer to enlist his participation as much as possible. In this respect, FSR amounts to being a preventive measure against rural landlessness. What promises it actually offers is unclear to its advocates who, as we saw earlier, have indicated uncertainty.

Underlying this uncertainty is a concern about the historical movement from traditional to modern agriculture. According to the team that reviewed farming systems research at particular institutes, traditional farming systems (that is, small-farm systems) must adjust to larger historical and social changes. The team stated that traditional systems were at one time disconnected to the larger national economy, which meant that these isolated systems were *optimal*. In the modern period, the team continued, the integration of once distant rural areas into the national economy has allowed these systems to 'become more open, more productive, more dynamic, more dependent upon purchased inputs and more vulnerable to changes in the environment'. In these terms, FSR must monitor this adjustment by having improved technology and farm practices incorporated into a farm system.

How traditional these targeted systems should or would remain stays unanswered. This apparent tension becomes even more glaring when debate occurs whether the small farmer is indeed deserving of as much attention as he/she has been receiving. The TAC quinquennial review of ICRISAT for example, exposed this debate. The review team questioned this institute's targeting of the small farmer and the consequent emphasis on zero of low-input farm conditions. The team, which conducted its review in 1978, argued that ICRISAT might be 'criticized for perpetuating agricultural under-development'.[47]

If agricultural underdevelopment is associated with small-farm low-input farming, the content of agricultural *development* is not discussed beyond the point of increasing farm productivity. But, by recalling our earlier discussion on the CGIAR's emphasis on distribution and small-farm development, the TAC statement on ICRISAT suggests research's contradictory interest in objectively promoting the rise of rural wage labour in the face of a stated interest in constraining rural proletarianization and its potential for social instability. We shall return to this argument. To move on, there are other programmes that parallel the development of FSR.

Removing constraints to higher yields: the insect pest problem

While FSR suggests to some that research may be promoting agricultural 'under-development' – low inputs, lower yields – and that it reflects a switch in research philosophy, international agricultural research has not relinquished its long-standing goal of increasing the yield per unit area of land. In this research system's view, FSR and increasing yields must be compatible. Central to getting down to the cultivator's world is the identification of the constraints which farmers face in increasing yields. New technology for the resource-poor farmer, it is assumed, will be profitable to the extent to which particular constraints are eliminated either agronomically or biologically.

Any observer will note that throughout the 1970s (and certainly in the 1980s) research reports ranging from yearly centre statements to CGIAR integrative reports to long-range plans underscore the theme of non-chemical control for the small, resource-poor farmer. There is no question that the introduction of improved, high-yielding plant varieties and their complementary inputs have complicated the insect pest problem. The combination of exogenously fed nitrogen fertilizer, irrigation and closer plant spacings created a favourable micro-climate for insect insurgence. Moreover, the new plant morphology (that is, short stature and stiff strawed) resulted in the transformation of previously minor and obscure pests into predominant ones – a process that has significantly altered the ecology of insects and rendered ineffective the performance of chemicals. This section will review one attempt to deal with these 'constraints to higher yields'.[48]

I have shown elsewhere that insecticidal applications were the mainstay of plant protection programmes during the early period of international agricultural research.[49] This situation gave way in the early 1970s to varietal resistance, which refers to plant breeding and entomological processes to incorporate insect-resistant genes into the plant. The unpredictable nature of the movement of insects along the minor-to-major continuum, the appearance of previously obscure and unknown diseases, and, of course, the constraining costs of insecticides rendered questionable the general effectiveness of chemicals. This, however, did not signal the end to chemicals. The idea of an 'all-weather' plant variety was (and still is) far-fetched. The vulnerability of *types* of resistance was shortly discovered in the farmers' fields.

The promise of varietal resistance was considerable. It reaffirmed the central role of plant breeding and genetics in Green Revolution research. But it soon gave way to an ensemble of approaches popularly known as integrated pest management (IPM). IPM includes biological control, varietal resistance, and certain agronomic/planting practices, the combination of which is designed to minimize insecticidal applications. As we show below, the connection between IPM and FSR is central to research's attempt to remove this constraint.

A deleterious consequence of varietal resistance for the resource-poor farmer is called varietal breakdown, which refers to the collapse of resistance levels. This result refers us to the saga of criticisms that pointed first to the farmer's dependence upon outside forces for plant material which he/she knows very little about; and secondly to the comparison of resistance levels of the so-called modern and traditional varieties.

The second charge continues to be a thorn in the side for this research system. The tone of an excerpt from a 1981 publication of an older centre indicates this:

> The modern ... varieties, sometimes called the HYV or high-yielding varieties, are often criticized because they are 'not as resistant to pests as the farmers' traditional varieties'. A favorite *ploy* of some critics is to tie the modern varieties to 'high-yields only where farmers use costly agricultural inputs'.[50] (Emphasis added.)

While controlled experimentation may show higher resistance levels in the greenhouse or laboratory, the tendency for varietal breakdown is a problem for research. Commenting on popular and quick experimental approaches to varietal resistance, the CGIAR's 1979 Integrative Report cautioned that 'breakdown in resistance can occur rapidly and in some instances replacement varieties may be required about every three years or so'.[51]

By way of rice research as an example, IRRI has found that the brown planthopper, which is the most notorious of rice insect pests, is a 'plastic species with a wide genetic variability. Because of its plasticity, it develops resistance to insecticides rapidly and over-comes the resistance of varieties through a selection process.'[52] This condition has led to the phenomenon of biotypes and the constant search for varieties resistant to specific types. IRRI, in 1981, had the following to say:

> Because of the genetic variability within the brown planthopper species, a small proportion of the individuals are able to feed on and damage a variety that is resistant to other individuals in the same population. When a resistant variety is planted over a wide area, the virulent insects that can survive and multiply on a resistant variety increase in number and become predominant after several generations. To keep ahead of the biotype selection process it is necessary to be able to evaluate breeding lines for resistance to a particular biotype before that biotype becomes predominant in the field. This way we can release varieties with genes for resistance to a particular biotype as soon as it is observed that the presently grown variety is becoming susceptible.

In the face of insect resurgence and plant breakdown, other breeding approaches have been summoned. The same Integrative Report noted an increasingly popular approach called multiline breeding, which refers to 'genotypes with same form and structure (phenotypes) but with differing sources of genetic resistance. This helps to ensure that if a new disease race appears, only a minority of plants will be susceptible', thereby preventing damage of epidemic proportions. CIMMYT and CIAT have apparently taken the lead in this area, with the former developing wheat multiline resistance to rust and the latter for what is known as blast disease.

A final and undeveloped approach is called horizontal or field resistance. This involves incorporating into varieties 'many minor genes that will provide a moderate level of resistance'. The CGIAR reported in 1979 that this technique has been recognized as best for long-term plant stability, but requires different screening methods for plant breeding. Said to be procedurally slow, horizontal resistance like biological control might actually fall within the basic research category having a longer time horizon.[53]

The struggle within science-based agricultural research will continue over the merits of each approach. But, as TAC described it in one of its quinquennial review reports, 'although there is still some academic argument among plant breeders concerning the merits of single-gene and polygenetic resistance to pests, history has shown that use of single-gene resistance has frequently resulted in short-lived effectiveness of resistant varieties'.[54]

Against the background of these debates, our point is that the promise of varietal resistance as a mainline, non-chemical approach had to be complemented in research's attempt to reduce risk and minimize farm inputs for the resource-poor farmer. To this end, IPM and its connection to FSR is central. Plant protection programmes at centres with established farming systems programmes have emphasized host plant resistance and agronomic approaches towards high and stable yields. CIAT, ICRISAT, and IITA are most explicit. The IITA's pest management programme is tied to its FSR land and soil management projects, 'with emphasis on maximizing cultural and biological control techniques and the minimum use of agro-chemicals'. CIAT reported to the Technical Advisory Committee in 1977 that 'when varietal resistance to diseases and pests cannot be found, control methods based on clean simple cultural practices and biological control are developed. Chemical control is only considered as a last resort or for the protection of clean planting material.'[55]

ICRISAT appears most impressive in its research practice of testing varieties under 'insecticide-free and low-fertility' conditions on its experimental station. The TAC review of the centre's cereal programme in 1979 stressed ICRISAT's small-farmer focus in its method of detecting 'the performance . . . of breeding material . . . under minimal farming practices'. The same situation applied to its pulse programme, most notably on chickpeas and pigeon peas, with *no* mention in the report of insecticidal treatments. The tie-in with FSR is through entomological and inter-cropping studies. These activities have included the study of population dynamics, pest monitoring through light-trapping and the survey and evaluation of biocontrol agents.

The earlier charges that cultivators could not afford chemical inputs and had to enter a state of indebtedness to purchase them were confronted by, first, recasting rural inequities in terms of technical constraints to higher yields and second, by proceeding technologically to eliminate them and rescue the resource-poor farmer from indebtedness. Here, we see the appearance of a sub-theme of the small-farm thrust in international agricultural research; that is, the importance of removing constraints for the small, resource-poor farmer. Thus, if the CGIAR, its centres and scientists see FSR as a particular approach to deal with the consequences of new technology, removing technical constraints is a corresponding method.

We must keep in mind, however, what constraints actually are. As defined, research projects investigating 'constraints to higher yields' have focused 'on that core of the constraints problem [on] which farmers and government agencies can exercise most control – the management of inputs and cultural practices'.[56] The targeting of the small farmer also embraces another costly input in nitrogen fertilizer.

Removing another constraint: increasing fertilizer efficiency and the prospect of genetic engineering

The 1973 'oil–energy crisis' resulted in an increased effort to improve fertilizer efficiency. These efforts were largely confined to agronomic–farming systems approaches that merged land management and soil fertility studies with investigations of plant responses to fertilizers. In addition, work took place on plant nutrients and the biological/genetic elimination (or reduction) of the fertilizer constraint. But the biological thrust has advanced of late, as a reading of recent CGIAR documents will indicate. In a word, the situation has developed owing to recent noted scientific advances and, connected to them, certain movements in the accumulation of private capital that have laid the objective basis for the entry of international agricultural research into this area.

Indicating the connection between small-farm development and lowering nitrogen fertilizer inputs are the discussions on bringing the International Fertilizer Development Centre (IFDC) into the CGIAR circuit of centres. Located in Muscle Shoals, Alabama, the

IFDC has been working with several of the international centres, notably IRRI by co-operating in the rice institute's International Network on Soil Fertility and Fertilizer Evaluation for Rice. At the request of the US Agency for International Development, the Technical Advisory Committee reviewed the IFDC for possible inclusion but did not recommend it in view of other priorities it (TAC) felt more urgent for agricultural research.[57]

This apparent secondary status of fertilizer and plant nutrients research was contested by two donor representatives at a May 1979 meeting of the Consultative Group. One speaker established the relationship of plant nutrient work and the small farmer. The representative argued that such research should be rated higher in view 'of an expert opinion that between 50 and 70 percent of nitrogen applied to rice paddies was lost and not absorbed by the plant'. Breakthroughs, this official continued, would be tremendous, especially for the 'smallest of the small farmers'. Along the same lines, a second speaker, presumably another donor representative, put forward a motion for a comprehensive review of all plant nutrients including inorganic fertilizer, compost, animal and human waste, and biological nitrogen fixation. 'He felt that a thorough examination of the whole plant nutrient question would lead the Group to pay more attention to it.'[58]

Within two years, the question deserved a systematic treatment in the Consultative Group's 1981 Integrative Report. The report stated that a 'search for biological means to enhance the fertility of agricultural soils through the action of micro-organisms' is imperative in view of 'the tightening of supply of fossil fuels and the rising costs of energy'.[59] The search, which refers to biological nitrogen fixation (BNF), would identify 'new crop lines that could take advantage of nitrogen fixation by algae and bacteria as well as more efficient nutrient uptake by root fungi'.

To be sure, BNF has long been part of world agriculture. According to the same report, rice farmers in South-East Asia have for centuries used the symbiotic connection of the water fern, *Azolla*, and blue-green algae as a source of plant nutrient. Used historically either as green manure before transplanting rice or as a rice intercrop, this technology has also been quite successful in China and Vietnam.

International research centres have various long-term studies under way. IITA, for example, has been experimenting with soybeans and cowpeas for nitrogen-fixing abilities in combination with symbiotic bacteria that supply nitrogen compounds through the plant root nodules. ICRISAT and CIAT have had similar programmes with pigeon/chickpeas and kidney beans respectively, while IRRI has investigated for several years free-living blue-green algae and *Azolla*. Thus, experiments have been conducted to identify and stimulate nitrogen-fixing ability of particular bacteria that are attracted to the root-zone of certain plants. And part of this research has also involved methods of inoculating the 'associative strains of bacteria into root-zones'.

Against this background, BNF-based fertilizer efficiency work is nothing new. The 1981 Integrative Report stated that until the mid-1970s, BNF research 'suffered a neglect'. The excitement, however, over the possibility of further plant breeding and biological work on this topic can be tied to two related developments. The first involves advancements and recognition of genetic engineering techniques, specifically protoplast fusion and direct injection of DNA.[60] For BNF, these developments could lead to sophisticated plant breeding work to transfer nitrogen fixating (nif) genes into major cereal crops.

The second is the link between genetic engineering as a form of high technology and the political–economic forces that drive it. The first half of this relationship is clear. It represents new areas of productive investment and, consequently, incentive for more scientific investigation. The same Integrative Report stated that the unknown consequences of genetic engineering could be likened to those of computers and electronics.

But this increased attention to plant breeding possibilities has developed along with the crisis in capital accumulation and with corporate investment shifting rapidly from oil-based, energy-intensive industry to 'hi-tech' sectors. In *Seeds of the Earth*, Pat Mooney revealed that 'in the last ten years, at least 30 seed companies with sales of $5 million or more have been acquired by large non-seed multinational corporate enterprises'.[61] While primarily profiting from pharmaceuticals or petroleum-based products, these corporate giants must control germ plasm for the successful commercialization of BNF-based technologies and bio-technology in general. Their names are familiar: Ciba-Geigy, Monsanto, Union Carbide, and Royal Dutch Shell.

While the connection between science-based research and capital accumulation is complex, we can predict that it will grow in complexity as dominant actors contend for the control of germ plasm. Our present point is simply this: that the objective base to further the biological imperative in international agricultural research is rapidly maturing in the area of plant nutrition and the removal of particular technical constraints to higher yields. It will certainly extend into the plant protection area, too. And it is maturing in light of the rapid motion of capital investment into hi-tech sectors, one of them being genetic engineering.

Expanding the circuit

Not only did the CGIAR, along with its centres, refine its research plan. It expanded the circuit of centres in the 1970s. Two additional organizations deserve our attention: the International Board for Plant Genetic Resources (IBPGR) and the International Service for National Agricultural Research (ISNAR). The establishment of both organizations corresponded to the needs and requirements of the Group's new agenda of technology for the small farmers.

To conserve genetic resources

It is not surprising that the Integrative Report for 1982 is devoted to genetic resource conservation following an extensive discussion the previous year on biological nitrogen fixation and genetic engineering. As we argued earlier, predominant actors have and will continue to scramble for germ plasm security owing to the requirements and promises afforded by hi-tech forms in agriculture. One of those actors is the Consultative Group through its centre devoted to germ plasm collection and storage – the International Board for Plant Genetic Resources.

The Board's establishment in 1974 was inspired by the history of genetic erosion and vulnerability, the tempo of which hastened during the Green Revolution period. In CGIAR's words, 'As the farmers shift to modern varieties they abandon their traditional varieties, many of which quickly disappear because their survival is entirely dependent upon cultivation by man. Unless grown and regenerated frequently, seed stocks lose viability and die, especially in the tropics and sub-tropics where many crops originated.'[62]

The IBPGR's work is consistent with the Group's small-farm strategy. The search for resistant genes is a never-ending one against the background of frequent varietal breakdown. Attention to horizontal resistance and/or field resistance discussed earlier necessitates the preservation of the so-called traditional plant varieties that are said to be more tolerant to a wider range of insects and diseases. Furthermore, FSR automatically points to more location-specific work in varietal improvement for drought-stricken areas, flooding, cold temperatures, and adverse soil conditions. The prevalence of these farm conditions brings forth the task of genetic preservation.

To these ends, the Board was established. Its mandate as found in its long-range plan for the 1980s reads as follows:

The innumerable varieties of crop plants of economic importance which have evolved throughout the world over some 10,000 years of domestication, comprise an irreplaceable reservoir of genetic material for future crop improvement purposes. But this genetic diversity has been disappearing rapidly as farmers replace a profusion of ancient land races with a relatively small number of new cultivars bred for high yield and other characteristics desirable in modern agricultural systems. At the same time, changes in land use and agricultural practices have been leading to the disappearance of the wild progenitors and weedy relatives of present-day cultivated plants which are also important for crop improvement. As a result of both of these developments, there is an urgent need to collect and conserve the diverse genetic plant materials which remain.

The basic mandate of the IBPGR is to help organize a global network of genetic resources centres whose activities will serve to safeguard, and make effectively available for future crop improvement purposes, the genetic variability of the major crop species, and of other plants of economic importance. The Board interprets this mandate as requiring it to encourage, promote and support the collection, conservation, characterization, documentation and utilization of germ plasm of all such crops and other plants.[63]

While inspired primarily by ecological and consequent economic conditions deleterious to farmers, however, the IBPGR's position can be questioned by recent movement towards BNF and genetic engineering. Clearly, the Board can anticipate an increasingly major role in international agricultural research as a contending actor in the germ plasm security area. But how will these developments (i.e. interests in hi-tech) affect the CG's overall small-farm bent? Whether or not the two interests are contradictory will depend on the political character and content of hi-tech activity. Robert S. Grossmann poses the question in the following terms: 'For whose benefits are genetic resources used? At whose costs are genetic resources unused?'[64]

Multinational capital's lead in the hi-tech arena and its primary interest in profits and capital accumulation, both of which involve the requirement to have genetic stocks that correspond to their specialized products, may result in the further narrowing of the genetic base. Strategic in this regard are long-term breeding goals that have considerable impact on all phases of the 'breeding pipeline', which begins with collection of germ plasm and ends with the production and distribution of high-quality seeds to farmers.[65] One of Grossmann's conclusions about corporate interests in this activity, especially its knowledge base, reads as follows:

Crop diversification (the introduction of under-exploited plants) might take place slowly, and changes in varieties used will continue to impact the labor force. New crops may creep in specialized niches, but they will be linked primarily to the processing industry. Crop vulnerability may also increase, even though the number of varieties may be increased, because the technologies will transfer some common genetic links among important cereal crops.[66]

Where the Consultative Group and its genetic conservation Board will position itself in the line-up of actors remains uncertain. If what Grossmann and others have argued is true, the Group and related actors will encounter serious contradictions that parallel those already expressed by the system's donors and centres in recent discussions reviewed in integrative reports. An indication of the line-up, though is evidenced in the maturing base for more bio-logical and genetic research in international agriculture. The base is maturing because of (or as a response to) multinational capital's general motion in genetic engineering and bio-technology. If this points to capital's ultimate control of these new areas, the inter-national circuit of donors and research centres will find it difficult to rid itself of contradictions in its programme.

To strengthen national agricultural research

In 1978, the donors of the Consultative Group agreed to establish the International Service for National Agricultural Research (ISNAR). The Service was a natural and logical outcome of research directions of the 1970s. The programmes and projects discussed above emphasize the location-specific character of research that targets the small farmer. FSR, IPM and, as an extension, BNF are based on understanding the economic–ecological setting of the resource-poor farmer before introducing new technology. It is obviously impossible for 13 international centres to cover all agro–climatic, ecological situations. In this context, strengthening national programmes becomes imperative for an international research system whose services depend on the capacity of national programmes to adapt farm technologies and practices generated at international centres.

The meaning of 'strengthening' national programmes is unclear. The question arose, certainly not for the first time, at a Group meeting at which donor representatives and centre directors dis-cussed a report by a task force that recommended the establishment of an ISNAR. Entitled *Report of the task force on international assistance for strengthening national agricultural research*, the task force recom-mended that ISNAR would assist national programmes on 'plan-ning, organization, and management issues'.[67] The Group's dis-cussion reflected the long-standing issue over the continuing presence or eventual absence of international research centres. An UNFAO representative called for a clearer definition of two types of activities: first, the strengthening of the capacity of the national programmes to absorb the technologies of international centres, which points to an explicit division of labour in a world research process and the continuing presence of international centres; or secondly, supporting and strengthening national programmes towards self-sufficiency and reliance, which suggests that international centres will eventually fade away.

Recalling for a moment an earlier period, the international centres were initially created to produce quick results and major breakthroughs and simultaneously strengthen national agricultural research programmes. In the course of history, the 'founding fathers' of international research speculated that the IRRI prototype would eventually become obsolete. A rather brief lifespan was anticipated.

We now know that the international centres have not withered away and are still with us attempting to cope with 'second and third generation problems'. The situation has been recast in CGIAR's second review: 'The consensus that is emerging, therefore, is that the International Centers will be needed to support national programs at least until the end of the century and that *they should be encouraged to evolve towards a system of institutions with integrated aims*' (emphasis added).[68]

The issue that UNFAO raised earlier is essentially rendered moot in the context of this consensus. Whatever the contention was in 1978, when the task force released its report, the issue was laid to rest in the Group's 1981 Integrative Report:

> By keeping abreast of new developments in science, the IARCs will be able to grasp the opportunities provided by new discoveries. The IARCs can continue to be innovators themselves by maintaining strategic research programs using up-to-date equipment and techniques. *The centers' commitment to strategic research with broad application will become even stronger as national programs begin to take over the site-specific, production-oriented applied or adaptive research. The IARCs will thus be able to devote more attention to difficult problems facing certain crop regions that require a protracted research effort.* The collections of crop germplasm that are scattered throughout the world can be exploited by the IARCs employing traditional breeding methods as well as new techniques; *there is an area where the centers have a clear comparative advantage over national programs, and where they can be of great service to the latter.*[69] (Emphasis added.)

Against this background, the establishment of ISNAR represents yet another stage of development for international agricultural research. If the emerging consensus seems to contradict the old philosophy (that of Harrar and Hill), it reaffirms and legitimizes, nevertheless, Harrar's vision of internationalizing the agricultural sciences to eradicate world hunger and poverty – a vision he expressed in 1954.[70] The current situation is thus more elaborate in its division of labour based on *integrated aims*. In this context, we can refer to *a* research system that is facing major contradictions that lie at the heart of its general policies and directions. We will conclude on this note.

Conclusion on agricultural research, the state and rural proletarianization

Central to our understanding of the current situation of international agricultural research are contradictions and inconsistencies in research policy that are based on an incomplete assessment of social conditions of capitalist agriculture in the Third World. Defining the widening social (class) polarities in terms of distribution and equity, research policy has targeted the small farmer, which, we argued, amounts to an attempt to resist the spread of landless, wage labour. Resistance represents research's effort to cope with the intensifying struggle within and between agricultural classes. In short, research has not been reacting solely to increasing landlessness, as if the landless were divorced from other classes. Its resistance is aimed at a potentially explosive political situation in which class interests collide.

As indicated earlier, policy contradictions were reflected in CGIAR discussions on the prospects of small-farm technology and on whether or not such technology would result in 'agricultural underdevelopment'. We may conclude from these discussions that the CGIAR and its review teams are not homogeneous; because there appears to be disagreement over key issues, we cannot validly point to *an* institution of international research. As a counter to this conclusion, our argument is that these contradictions and inconsistencies are not happenstance. Instead, they are unresolvable and represent a more general position towards rural proletarianization under increasing immiseration. I am referring to research's position of neutrality.

This position takes several forms. One is that international agricultural research is developing technology and no more. Another, which we will soon discuss, is that the state, and not research, is responsible for dealing with the problem of distribution. And a third echoes debates within the CGIAR: it is virtually impossible to develop technology for the small farmer because large

farmers will also find smallness advantageous, and therefore proletarianization is inevitable. But the overall position points to the
role of the state in agriculture and research's 'neutral' relationship to
the state. Ultimately, the contradictions outlined throughout this
chapter are unresolvable because of this relationship and, consequently, the location of research in the polarization of agricultural
classes. In other words, research must take a position of neutrality
precisely because it is not neutral.

The basis for bringing the state, or government in the CGIAR's
language, into our analysis is that the circuit has considered it
important in several recent reports. Research articulates its
neutrality in two respects. The first attempts to reaffirm its international, non-governmental character. The second has to do with
the consequences of its work.

The CGIAR's second review committee underscored the circuit's
preference for autonomy. After reviewing research's relationship
with other institutions (scientific and financial/governmental) in the
field of international agriculture, the committee concluded that the
CG donors ought to restrain themselves 'to avoid attempts to
influence the policies of an Institution without regard to the policies
of the Group'. The committee described the current situation in
terms of international centres sitting on a 'thinly drawn line between
autonomy and dependency'. And on this line, it contended, are
three hypothetical structures:

(1) The Institutions could be viewed as the product and responsi-
     bility of the international community of nations as a whole.
     The nature of the Group would then change to become a
     governmental organization similar to those of the United
     Nations system.
(2) The Institutions could be viewed as research organizations that
     not only work for, but are part of, the institutional framework
     in the developing countries. The Group would be seen as
     supporting the institutions as a way of helping the developing
     countries to evolve regional research institutions of their own.
     Funding by donors would be seen as an interim measure,
     decreasing with time, until the countries themselves funded the
     institutions on an inter-governmental basis.
(3) The Institutions could be viewed as international and non-
     governmental, by allowing priorities to be determined largely
     by individuals appointed for their own personal, professional
     ability. The institutions would be funded through the Group
     on a long-term basis, by donors who would be free to continue
     or withdraw their support as they saw fit. The international

character of the system would stem from the support of the Co-sponsors as well as from the composition of the Group, the Boards and the Technical Advisory Committee.[71]

In the end, the committee reaffirmed the system's international character expressed in the third structure, which also described the predominant structure that began with the establishment of the International Rice Research Institute. The first structure was discounted with little discussion. It seemed from the report's content that more time was spent on the second, the discussion of which clearly reflected the circuit's general position of neutrality. The second structure, the committee said:

> clearly implies that Institutions supported initially by the CGIAR would progressively move towards becoming regional intergovernmental organizations with greater funding by developing countries. It would also probably introduce political considerations into the deliberations of Boards and hence lead to the instability of funding that has characterized regional institutions in the developing world. In addition, political considerations would also become more important in the policy decisions of the Group because budgetary decisions would be more closely associated with specific developing regions. The logical system of organization and management would stress the donor nature of the CGIAR and the dominance of developing country representation on the Boards. The donors would exercise the ultimate power of sanction through funding policies. Discussion and negotiation would be between the Group as donors and the Centre Boards as representatives of beneficiary countries.[72]

The system's desire for political autonomy and neutrality is clear, as is its aversion to being incorporated into a governmental/institutional framework that would burden it with 'political considerations'. The questionable nature of this assessment is not based on the degree of donor influence upon the present situation. Instead, as we argue below, it is based on research's ultimate and highly political position on the consequences of its work.

The second review of the CGIAR reveals the circuit's current and perhaps final position on the consequences of new technology. The review committee stated the following:

> First, technology is only one of the elements that determine agricultural production and productivity. Furthermore, the adoption of technology is the result of complex socio-economic and political processes where research and research institutions may have only a relatively small role. Consequently, the magnitude of the impact of new technology on production and productivity must be viewed within this larger framework of social and political phenomena.

Second, the impact of a specific technology on income distribution and the well-being of small farmers is only partly determined by its inherent characteristics (biological, mechanical, etc.): it is determined to a far greater extent by the institutional framework within which the productive processes take place. Under different economic and institutional frameworks the same technologies will have different impacts on income distribution and on the well-being of the small, poor farmer . . .

Nevertheless, an analysis of research and technical change must take these elements into consideration. All concerned need to take a balanced view and to make an effort to understand the impact of technical change *within the wider framework of the socio-economic and political forces that both determine and are affected by it.*[73] (Emphasis added.)

It is impressive that this research system has expressed its position in these terms. International agricultural research policy reviews and statements failed to consider, just a decade ago, the 'wider framework of the socio-economic and political forces that both determine and are affected by' research and technological development. Like many of the recent CGIAR and centre formulations of the consequences problem, the passage above reaffirms and restates criticisms voiced in the early 1970s.

But the desire for neutrality remains paramount. Despite the recognition that research and development are an identifiable component within the wider framework, international agricultural research justifies itself according to terms separate from it. In the Consultative Group's terms, research is simply one element in the strategy to improve world food and agriculture and, therefore, agricultural research cannot be fully accountable for the consequences of its work that are reserved for the wider framework of social forces. This becomes especially clear, for example, in statements that consistently stipulate that technology alone cannot solve the world's food problem; or, that it 'is a necessary but by no means sufficient component of a world food strategy.'[74]

In this context, the Group and its centres have been consistent about who stands to benefit from technical change. It is true that donors have encouraged the centres to target the small, resource-poor farmer. But in its 1979 'priorities' paper, the CGIAR's Technical Advisory Committee recommended that:

due account be also taken of the need to achieve an improvement in the level of income and standard of living of the less advantaged sectors of society in the developing countries (especially rural), which determine their access to food, equity in distribution of benefits from research, and efficiency in the use of agricultural resources.[75] (Emphasis in original.)

The meaning of 'due account' is unclear. It may allude to historical neglect, which ought to be rectified or compensated for.

On the other hand, 'due account' can also reflect a *conditional* (and neutral) concern on the part of research. The second interpretation takes on increasing significance when we consider statements that explicitly caution research's involvement in this area. With enough evidence collected about the political–economic relationships that embrace technology, TAC also stipulated that 'the benefits derived from international agricultural research by different social groups would however very much *depend on the conditions of the country concerned and are a matter of consideration by the individual governments in establishing their development policies and plans*' (emphasis added).[76]

What, then, are these conditions? Our answer can begin by recapitulating the argument that international agricultural research's programme has been based on a partial, incomplete assessment of social conditions. By defining its tasks in terms of the static, once-and-for-all concepts of distribution and equity, it has nullified the essence of the earlier criticisms which stressed the social relations of production and relations of power, which the rise in proletarianization has represented.

Similarly, TAC's *open-ended* referral to 'conditions' repeats the partial treatment of agriculture outlined above. In pointing to the state, research's partiality essentially corresponds to the state's history of 'inaction' towards resolving contradictions in the social organization of agricultural production. On the contrary, state policy has taken to controlling, monitoring, and reacting to rising social tensions that are symptomatic of the tension between classes. Inaction in this political situation is displayed by the state's interest in controlling the pace of proletarianization when conditions become unbearingly unstable, while *at the same moment* endorsing and promoting it to enforce the reproduction of the dominant class structure which the state, finally, represents and defends.[77]

The objective correspondence between programmes of the state and international agricultural research becomes apparent. Research, too, has taken to responding to the antagonisms within and between classes through its small-farm project. And in doing so, research and its preference for political neutrality, reaffirm the state's requirement for legitimacy and reproduction.

Faced with this internal contradiction, however, research cannot afford to be quiescent. Tied to the institutional framework of the state, research must offer 'neutral' answers. Indicative is the more recent and aggressive approach of international agricultural research towards agricultural policy, demonstrated by the inclusion into the circuit of the International Food Policy Research Institute (IFPRI). The second review committee report informs us that 'the inclusion of IFPRI in the System is related to the better understanding that has evolved in recent years of the interdependence between technical change in agriculture and socio-economic processes'. The report also suggested that IFPRI is expected to extend the system's understanding of small-farm behaviour to cover 'those of policy-makers themselves as determinants of technical change'.[78]

Entry into this area is not only confined to IFPRI. The 'elders' of the system, CIMMYT and IRRI, have started seminars specially for policy makers. In 1977, CIMMYT's Board of Trustees recommended a three-year plan for such seminars. At its annual International Rice Research Conference in 1982, IRRI's director-general discussed the role of policy makers in small-farm development.[79]

As a final question, what does international agricultural research have in store for the policy arena? A presentation by IFPRI's director, John Mellor, at a 1979 meeting of Group donors and other centre directors, provides some hints. While calling 'the research process ... really the basic engine of the growth process in agriculture', Mellor's testimony drew attention to the labour situation and the need to increase employment opportunities, which are required to provide 'effective demand' for surplus food production owing to the introduction of new agricultural technologies. To be sure, Mellor cautioned, 'the unseen hand of Adam Smith' cannot fully apply in this situation. In addition to Smith's invisible hand, 'we [IFPRI] want to see how much of the process and of the problem can be dealt with through these kinds of processes; to know what kinds of governmental policies are needed to encourage them; and, what kinds of supplemental policies may be needed to make up for deficiences in the process which we are describing'.[80]

The importance of labour to IFPRI's formulation suggests that this research system, first, expects proletarianization to continue, which is a position that reaffirms contradictions in the small-farm thrust; and, second, that this social and potentially political motion has to be dealt with. Policy options, however, remain problematic. It seems that these options will take the general form of shifts in state-governmental budgets to favour the rural/agricultural sector so as to create 'a sort of agriculture-led strategy' based on small-farm technology and the generation of employment for the rural poor.

Proposals and measures like these leave suspended questions about the endemic crisis of capital accumulation in nearly all Third World countries, which has much to do with the (un)likelihood of removing budgetary constraints and developing favourable 'five-year' plans. Also not dealt with is the state's related inability to drain enough surplus capital out of agriculture for investment in other sectors to create employment opportunities.

If the state is unable to generate surplus capital necessary to accommodate an expanding wage labour force, there can be no other consequence but increasing immiseration and, very likely, the continuing repression of labour that prevails in these countries. As a general tendency, it will further result in rising social tensions, which certain sections within the research system have been predicting. Recalling Byres' essay on the Indian countryside:

> If labor is 'trapped', it is a trap that seems likely to become increasingly intolerable if, with the prospect of a diminution in the level of employment, rich peasants are successful in acquiring a yet greater proportion of the operated area and try to force the pace of proletarianization. The outcome will be determined in the arena of class struggle, in a possible struggle over land.[81]

To conclude, a partial analysis of social conditions, the resultant programme for the small farmer, and the general position of neutrality locate research in the institutional framework of the state. This conclusion is based on the harmonious and reaffirming relationship between research and state programmes for capitalist agriculture. Within this framework and the historical context of widening social polarities, international agricultural research may be objectively saying that proletarianization must take on its own independence and historical course.

Notes

This paper was completed in the autumn of 1982 at the International Institute for Environment and Society of the Science Center, Berlin. I would like to thank the Institute, especially Dr Bernhard Glaeser, for providing a generous grant and excellent working conditions during my stay in Berlin.

1   First, a few introductory comments. I have taken the liberty of using throughout this chapter the general term 'research' or 'international agricultural research' to refer to our subject. I feel we can discuss this topic in terms of a system (or circuit) that represents a general institution of research. The reader may also be disturbed at the liberty I have taken to lump together donors, centres, and scientists as constituting a research system, when in fact there are disagreements over particular issues among these units of the system. But I would maintain that these disagreements do not negate the political and strategic importance of focusing on a system. Finally, I wish to warn the reader that this chapter will not cover the entire range of research programmes that are funded by the CGIAR. I will select information on those centres whose work shows an immediate connection to the early criticisms.

2   See World Bank (1981), p. 40.

3   Ibid., p. 13. See also CGIAR (1981), pp. ix, 33, especially Ch. 3.
4   Mooney (1980), p. 38.
5   In Harrar (1967).
6   See CGIAR (1977), p. 46.
7   Oram (1982), p. 10.
8   CGIAR (1977), p. iii.
9   CGIAR (1981), p. 24.
10   CGIAR (1977), p. iii.
11   Letter from Warren Baum to CGIAR members, 22 September 1981, in CGIAR (1981).
12   CGIAR (1981), p. xiii.
13   Ibid., p. 32.
14   CGIAR (1977), p. 35.
15   The 'resource-poor farmer' is the expression used in CGIAR circles.
16   Some of the UNRISD studies are Griffin (1972), Palmer (1974 & 1975), Hewitt de Alcantara (1976).
17   Pearse (1977), pp. 126–7.
18   Ibid., p. 147.
19   Byres (1981), p. 427.
20   Ibid., pp. 430–1.
21   Ibid., p. 430; Byres, in presenting this argument, relied on earlier empirical studies, e.g. Rao (1975) and Bardhan (1976). Regarding the adoption problem, Byres notes that 'A sharecropping poor peasant who "adopts" the new inputs through cost-sharing leasing and who pays an increased share to his landlord is very different from a rich peasant who adopts the new technology' (p. 427).
22   See literature of dependency theory.
23   Collins (1981).
24   Mooney (1980), p. 11.
25   Dahlberg (1979), p. 81.
26   Ibid., p. 82.
27   Ibid., pp. 83–4.
28   Why IRRI and not CIMMYT appears to be due to the overshadowing presence of plant breeding and varietal improvement at CIMMYT. One indication of this is found in a report of a review team assembled by the CGIAR's Technical Advisory Committee (TAC), which quinquennially reviews the system's centres. The TAC team found that 'It has not been an easy task to bring economics discipline to bear on programs headed by biological scientists who have felt they could provide needed judgement of an economic character.' See TAC (1976), pp. 48, 63. TAC is a group of a dozen scientists prominent in international agricultural research and development. Its task is periodically to review each centre as well as to assess priorities for international agricultural research.
29   International Rice Research Institute (1970), p. 9. I have documented this crucial period in considerable detail in my Ph.D. dissertation (Oasa 1981).
30   The survey, later published under the title *Changes in rice farming in*

selected areas of Asia, sought to answer questions about adoption rates and levels, and adoption constraints, employment consequences, and distribution of benefits. The last was phrased as follows: 'Who has benefited from the new technology, and how have these benefits been spent? To what degree are profits capitalized into rising land values? How has the relationship among various tenure groups (landowners, tenants, and landless labor) changed?' Some of the *Changes* reports stated that the technology tended to benefit landowners, 'and the unfavourable position of small landowners and tenants in the inputs, credit, and factor markets have increased inequalities'. These same reports also warned that the new technology could potentially be 'seeds of future tension and conflict' in the countryside. Statements like these, we will find, became more frequent throughout the CGIAR systems and, most certainly at the time, could only impel IRRI to investigate the matter more thoroughly.

31   IRRI, Department of Agricultural Economics (1976), p. 1.
32   IRRI, Department of Agricultural Economics (1978), p. 6.
33   Binswanger *et al.* (1978), p. 408.
34   CGIAR Secretariat (1977), pp. 5, 26.
35   CGIAR Secretariat (1978), p. 13.
36   Ibid., p. 13.
37   Excerpts from the CGIAR Secretariat, informal summary of proceedings of International Center's Week and Consultative Group meeting. Washington, DC, 28 February 1979, pp. 12–13.
38   TAC (1978a), p. 8.
39   Ibid., p. 14. 'IARCs' stands for the International Agricultural Research Centers.
40   See, for example, IRRI (1970).
41   Brady *et al.* (1973).
42   TAC (1978a), Annex 2, p. 3.
43   The programme was disbanded in 1975 because the task was judged exceedingly complex owing to the region's diverse social and ecological conditions.
44   This was reported in TAC (1977), p. 5.
45   Ibid.
46   TAC (1978b), p. 12.
47   TAC (1978a), pp. 13–14.
48   Some of the international Centres have had what are called 'constraints' studies for a number of years. CGIAR Integrative Reports have highlighted these studies as integral to the system's efforts to deal with the social consequences of new technology. In fact, IRRI formalized the constraints approach. According to IRRI's Department of Agricultural Economics, its Constraints-to-Higher-Yields Project has focused on that core of the constraints problem over which farmers and government agencies can exercise the most control – the management of inputs and agronomic practices. See, for example, Barker (1978).
49   See my dissertation (Oasa 1981). To be sure, this situation did not mean that non-chemical approaches were not tested. In the mid-1960s

John L. Nickel, IRRI's entomologist at the time and now director-general at CIAT, spent two years researching the manipulation of natural enemies to reduce economic losses from pest species. Nickel completed a manual on the topic and encouraged experimentation with imported parasites and ecological surveys. Nickel's work was not pursued and chemicals remained predominant (see Nickel 1967, p. 443; also Nickel 1964).

50   IRRI (1982a), p. 20.
51   CGIAR Secretariat (1979), p. 13. In this passage, the CG was referring to rice technology and what is known as vertical resistance, which is popular because it is quicker to develop in experimentation. Vertical resistance depends upon the inbreeding of a major gene (or the stacking of major genes) known to possess strong resistance to particular insects.
52   IRRI (1982a), p. 23.
53   Related to field resistance is the notion of tolerance, whereby a variety is able to grow and flower despite a high insect population.
54   TAC (1978b), p. 65.
55   Ibid., p. 27; and TAC (1977), p. 17.
56   See Barker (1978).
57   TAC (1979a & b).
58   CGIAR Secretariat, informal summary of proceedings of Consultative Group meeting. Paris, 3–4 May 1979, p. 5.
59   CGIAR Secretariat (1981), p. 2.
60   According to ibid., protoplast fusion was described as follows: 'cells from different plants are placed together in a medium with a chemical which encourages clumping of the cells. An enzyme is then added to the medium to destroy the cell walls, thus allowing the cellular contents to merge. Genetic material can thus be exchanged between different species and even different genera.'
61   Mooney (1980), p. 55.
62   CGIAR Secretariat (1982), p. 2. According to Mooney (1980, p. 3), the great historical centres of diversity are located in less than one-quarter of the Earth's arable land. Based on the conclusions of a Russian scientist of the 1920s, the major areas are the Mediterranean, the Near East, Afghanistan, India–Burma, Malaysia–Java, China, Guatemala–Mexico, the Peruvian Andes, and Ethiopia.
63   International Board for Plant Genetic Resources (1981), p. 1. The global network alluded to in the mandate includes national programmes, most notably the National Seed Storage Laboratory at Fort Collins, Colorado, and international agricultural research centres. According to Mooney (1980), there are approximately 60 national programmes throughout the world.
64   Grossmann (1982).
65   Other phases that Grossmann (1982) listed were as follows: 'understanding the genetic variability and geographic distribution of cultivated species and their taxonomic and cytological relationships with closely related species; screening plant genetic resources for specific desirable

characteristics; studying the genetic mechanisms controlling the inheritance of desirable characteristics; combining genes from diverse sources into improved strains more useful to plant breeders; and breeding, releasing, and maintaining breeder seeds of varieties and stocks of improved germplasm.'

66 Grossmann (1982), p. 25. Two other conclusions forwarded by Grossmann are, first: 'Private industry will probably continue to believe it is in their interest to allow the land grant system to focus on pre-breeding for such traits as pest and disease resistance. Industry will be more interested in capturing genetic innovations (e.g. nitrogen-fixing corn) that will expand their markets'; and secondly: 'It is questionable whether a long-term public interest is pursued through such arrangements. This conclusion appears even more compelling if one considers who carries what kind of risks. An episode similar to the corn blight does cause some economic problems for the seed industry. Given patterns of integration among American corporations with agricultural interests, the losses may not be all that substantial. Farmers lose through lost income. Taxpayers lose through bailout schemes. Consumers lose through higher prices. If the public agricultural research system takes responsibility, who pays the bills?'

67 CGIAR (1978), p. ii.
68 CGIAR (1981), p. 32.
69 CGIAR Secretariat (1981), p. 19.
70 Harrar (1967).
71 CGIAR (1981), p. 37–8.
72 Ibid., p. 38.
73 Ibid., p. 15.
74 Centro Internacional de Agricultura Tropical (1981), p. 13.
75 TAC (1979b), p. 7.
76 Ibid., p. 11.
77 The notion of 'defence' comes from Offe and Ronge (1982), pp. 249–56. The notion of 'correspondence' is derived from Poulantzas (1973, 1975).
78 CGIAR (1981), p. 51.
79 IRRI (1982b), p. 15. M. J. Swaminathan, IRRI's director-general, said: 'Hence, it is important that policy-makers also become aware of the *constraints* responsible for the gap between partial and actual farm yields. IRRI should periodically organize seminars for policy-makers, based on carefully conducted case studies.'
80 Statement on IFPRI by John Mellor at the International Center's Week and Consultative Group Meeting, 29 October–1 November 1979, p. 100.
81 Byres (1981), p. 435.

References

Bardhan, P. 1976. Varieties in extent and forms of agricultural tenancy. *Economic and Political Weekly* **37** & **38** (11, 18 September).

Barker, R. 1978. *Adoption and production impact of new rice technology – the yield constraints problem.* Conference on Farm Level Rice Yield Constraints, IRRI, Los Banos, 24–6 April.

Binswanger, H. P., V. W. Ruttan and others 1978. *Induced innovation: technology, institutions, and development.* Baltimore: Johns Hopkins University Press.

Brady, N. C., D. S. Athwal and F. F. Hill 1973. *A proposal for broadening the mission of the International Rice Research Institute.* Los Banos: IRRI (mimeo).

Byres, T. J. 1981. The new technology. Class formation, and class action in the Indian countryside. *J. Peasant Studies* **8**.

Centro Internacional de Agricultura Tropical 1981. *CIAT in the 1980s.* Cali, Colombia: CIAT.

CGIAR 1977. *Report of the Review Committee.* Washington, DC: CGIAR.

CGIAR 1978. *Report of the task force on international assistance for strengthening national agricultural research.* Washington, DC: CGIAR.

CGIAR 1981. *Second review of the CGIAR.* Washington, DC: CGIAR.

CGIAR Secretariat 1977. *The Consultative Group and the international agricultural research system: an integrative report.* Washington, DC: CGIAR.

CGIAR Secretariat 1978. *The Consultative Group and the international agricultural research system: an integrative report.* Washington, DC: CGIAR.

CGIAR Secretariat 1979. *1979 Report on the Consultative Group and the international agricultural research system: an integrative report.* Washington, DC: CGIAR.

CGIAR Secretariat 1981. *1981 Report on the Consultative Group and the international agricultural research it supports: an integrative report.* Washington, DC: CGIAR.

CGIAR Secretariat 1982. *1982 Integrative report.* Washington, DC: CGIAR.

Collins, J. 1981. *On the state of global food and agriculture.* Symp. on Critical Perspectives on Global Food and Agriculture, Honolulu, Hawaii.

Dahlberg, K. A. 1979. *Beyond the Green Revolution: the ecology and politics of global agricultural development.* New York: Plenum Press.

Griffin, K. 1972. *The Green Revolution: an economic analysis.* Geneva: UNRISD.

Grossmann, R. S. 1982. *Control of agricultural germplasm and genetic information.* Conference on Biotechnology, Agriculture and the Public Interest. University of Kentucky, Lexington, 16–18 June.

Harrar, J. G. 1967. *Strategy toward the conquest of hunger.* New York: The Rockefeller Foundation.

Hewitt de Alcantara, C. 1976. *Modernizing Mexican agriculture: socioeconomic implications of technological change, 1940–1970.* Geneva: UNRISD.

International Board for Plant Genetic Resources 1981. *The IBPGR in the eighties: a strategy and planning report.* Rome: IBPGR Secretariat.

International Rice Research Institute 1970. *Rice research and training in the 70's*. Los Banos: IRRI.

International Rice Research Institute 1982a. *Research highlights for 1981*. Los Banos: IRRI.

International Rice Research Institute 1982b. *Rice research in the 1980s: summary reports from the 1982 International Rice Research Conference*. Los Banos: IRRI.

International Rice Research Institute, Dept of Agricultural Economics 1976. *Economic consequences of new rice technology: a view from IRRI (a proposal for a conference)*. Los Banos: IRRI (mimeo).

International Rice Research Institute, Dept of Agricultural Economics 1978. *Consequences of new rice technology*. IRRI Internal Annual Review, 25 January. Los Banos: IRRI.

Mooney, P. R. 1980. *Seeds of the Earth: a private or public resource*. Ottawa: Canadian Council for International Cooperation.

Nickel, J. L. 1964. *Biological control of rice stem borers: a feasibility study*. IRRI Technical Bull., no. 2. Los Banos: IRRI.

Nickel, J. L. 1967. The possible role of biotic factors in an integrated program for rice stem borer control. In *Symposium on the major insect pests of the rice plant*, ed. IRRI. Baltimore: Johns Hopkins University Press.

Oasa, E. K. 1981. *The International Rice Research Institute and the Green Revolution: a case study on the politics of agricultural research*. University of Hawaii, Honolulu.

Offe, C. and V. Ronge 1982. Theses on the theory of the state. In *Classes, power, and conflict: classical and contemporary debates*, A. Giddens and D. Held (eds). Berkeley: University of California Press.

Oram, P. 1982. *Strengthening agricultural research in the developing countries: progress and problems in the 1970s*. Paper for the Consultative Group on International Agricultural Research, Paris, 24–6 May.

Palmer, I. 1974. *The new rice in monsoon Asia*. Genevia: UNRISD.

Palmer, I. 1975 *The new rice in the Philippines*. Geneva: UNRISD.

Pearse, A. 1977. Technology and peasant production: reflections on a global study. *Development and Change* **8**.

Poulantzas, N. 1973. *Political power and social classes*. London: NLB.

Poulantzas, N. 1975. *Classes in contemporary capitalism*. London: NLB.

Rao, C. H. 1975. *Technological change and distribution of gains in Indian agriculture*. Delhi: Macmillan of India.

Technical Advisory Committee 1976 *Report on the TAC quinquennial review mission to the International Maize and Wheat Improvement Center*. Rome: TAC Secretariat, UN Food and Agriculture Organization.

Technical Advisory Committee 1977. *Report of the TAC quinquennial review mission to the International Center of Tropical Agriculture*. Rome: TAC Secretariat, UN Food and Agriculture Organization.

Technical Advisory Committee 1978a. *Farming systems research at the international agricultural research centers*. Washington, DC: CGIAR.

Technical Advisory Committee 1978b. *Report of the TAC quinquennial review mission to the International Institute for Tropical Agriculture*. Rome: TAC Secretariat, UN Food and Agriculture Organization.

Technical Advisory Committee 1979a. *TAC conclusions and recommendations on the inclusion of the International Fertilizer Development Centre (IFDC) in the CGIAR system.* Rome: TAC Secretariat.

Technical Advisory Committee 1979b. *TAC review of the priorities for international support to agricultural research.* Rome: TAC Secretariat.

World Bank 1981. *Agricultural research: Sector policy paper.* Washington, DC: World Bank.

3 The Green Revolution re-examined in India

PIERRE SPITZ

Political labelling of an agricultural strategy and early debates

As early as December 1969, the Green Revolution was discussed before the US House of Representatives at the Subcommittee on National Security Policy and Scientific Development of the Committee on Foreign Affairs. The title given to the publication of the proceedings was *Symposium on science and foreign policy: the Green Revolution*.[1]

During the discussions, the Green Revolution was presented as a major tool of United States foreign policy, even its most important element according to Charles S. Dennison, Vice-President of International Minerals and Chemicals Corporation, who was enthusiastic about the prospects of the growth of the fertilizer industry. Representative D. H. Fraser (from Minnesota, the 'Wheat State') stated that the term 'Green Revolution' was used for the first time by William S. Gaud, the US AID administrator, in a speech before the Society for International Development in March 1968. Whoever coined the phrase has made a tremendous job of public relations. It is still in use in the 1980s – a remarkable achievement, considering the rapid erosion to which the vocabulary of development policies is usually subjected. 'Green', of course, was implicitly opposed to 'red', and was signalling, like a flag, that social reform was not necessary, since technical means in agriculture (evoked by 'green') alone were supposed to solve the problem of hunger. The advocates of social reform reacted very quickly. The FAO review, *Ceres*, provided them with a platform to voice their warnings about the social impact of the Green Revolution: Edmundo Flores (*Ceres*, May–June 1969) and Solon Barraclough (*Ceres*, November–December 1969) wrote forcefully on this subject.

Along similar lines, Wolf Ladejinsky, working for the Ford Foundation and later for the World Bank, published in 1969 in the *Economic and Political Weekly* (Bombay) two articles on the Green

Revolution in Punjab[2] and in Bihar.[3] Ladejinsky stated that in
Punjab the new wheat technology was widening the gap between
the rich and the poor and pointed out that in Bihar the prospects of
rice farmers were more restricted than those of wheat farmers and
that only a small number of them would take advantage of the new
seeds. In 1970 he wrote an article in *Foreign Affairs* (July 1970) on the
'Ironies of India's Green Revolution', further stressing the contradictions of the new technology and warning of potential social
conflict – an indirect answer to the eulogy written two years earlier
in this influential American journal by Lester Brown, who later
changed his views.

Since 1970, the year the Nobel Peace Prize was awarded to
Norman Borlaug, the debate has continued between supporters and
critics of the Green Revolution. A major research project, financed
by the United Nations Development Programme and carried out by
the United Nations Research Institute for Social Development
(UNRISD) under the leadership of Andrew Pearse, refined and
substantiated early criticisms on 'the social and economic implications of a large-scale introduction of new varieties of food grain'.
Some 15 volumes were published and contain detailed analyses of
agricultural and rural changes.[4] After this research programme was
completed, UNRISD embarked on a new research programme
entitled 'Food systems and society', first developed in India, then in
Mexico and Nicaragua and more recently extended to Senegal,
Upper Volta, the Ivory Coast, Chile, Bangladesh and China.

This contribution, drawing on a draft interim report on India[5]
written within the framework of the 'Food systems and society'
project, reviews some of the features of the Indian food system
which call for more attention than in the past: the imbalance
between resources devoted to wheat and rice, with a preference
given to wheat, against other food grains; the neglect of rain-fed
areas; the neglect of small farmers and agricultural labourers, and
more generally, of issues of employment and purchasing power and
their seasonal aspects; the preference given to homogeneity in
cropping systems and the neglect of mixed-cropping systems; the
interactions between agricultural research priorities and social forces
(the 'participation' issue), and between technological choices and
environmental conservation (the 'sustainability' issue).

Instead of merely looking at the consequences of technological
choices made in the past, we should plead for alternative technologies which could allow a less uneven development between crops,
regions and social groups – reducing disparities between them as
well as between seasonal income, employment and year-to-year
production, while ensuring in the long term a sustainable develop-

ment by carefully avoiding the degradation of natural resources on which agriculture is based. A different set of technological choices, however, such as those indicated by I. Sachs (Ch. 9 below), need social forces to support them. At the national level, a different organization of agricultural research, education and extension is required, allowing farmers' know-how and creativity to be released through two-way channels of participation with scientists and agricultural extension agents. In turn, this requires the recognition of people's potential creativity. The expression 'Green Revolution' is a major ideological obstacle to this recognition, as it implies breaking with old farming systems and techniques, instead of submitting them and the vernacular knowledge crystallized in them to the most sophisticated scientific research. Prisoners of the history of their own agriculture and agricultural science as well as their short-term economic and political interests, industrialized countries cannot be expected to reverse the trend. They continue through their academic institutions and funding priorities to support an agricultural model which, whatever increases can be obtained at the national food output level, cannot solve the problems of total or seasonal unemployment and in so doing eradicate hunger. Policies of the Consultative Group on International Agricultural Research (CGIAR) are, as demonstrated by Edmund Oasa (Ch. 2, above), a reflection of this international state of affairs. It is therefore the task of concerned citizens, scientists and policy makers to put pressure on their governments and institutions to foster social and technological change in order to improve the food security of all social groups throughout the seasons and years.

Rice: slow progress

In India the area under food grains (around 125 million hectares) makes up about three-quarters of the total cultivated area and is divided into three almost equal parts: rice, wheat plus pulses, and coarse cereals (sorghum, millet, and so on), the 1980 percentages being 31.5%, 35.2% and 33.3% respectively.

Paddy is the most important of the food grains by area sown (31.5%) and by production (about 38.8% in 1980). The high temperature and water requirements of paddy make it predomin-antly a crop of the deltas and floodplains. It is mainly cultivated without irrigation in areas of high precipitation or low rainfall variability, and under irrigated conditions in areas of light annual precipitation or high rainfall variability. As a result of these climatic constraints, the major rice-producing states are West Bengal, Tamil

Nadu, Andhra Pradesh, Bihar, Orissa and Uttar Pradesh. Each of these states produces more than 4 million tons of rice and together they account for two-thirds of the country's rice output.[6] Three of the major rice-producing states are in eastern India (Bihar, Orissa and West Bengal), and to these should be added the eastern districts of Madhya Pradesh (with an output of more than 3 million tons). The Punjab is also emerging as an important rice producer, with more than 3 million tons. Rice has become a second crop in this Green Revolution area, a trend which can also be seen in Haryana and western Uttar Pradesh.

North of the 17° latitude, paddy is, as a rule, sown after the rains have begun (June–July) and harvested in November–January. This wet season is generally referred to as *kharif*, and paddy is typically a *kharif* crop. In eastern India, irrigation is used to supplement rainfall during the wet season and to allow production of one or two dry season (*rabi*) crops. Such irrigation is generally canal irrigation, the water flooding from field to field with little control by farmers. Once water has reached the main channel, the outlets are not usually closed, so that water flows continuously through the fields, resulting in very uneven distribution. In fact, the flow may not always reach the fields at the end of the service area or at the end of the canal.

In West Bengal, the *kharif* crop is locally known as *aman*, and is the principal crop, accounting for three-quarters of the acreage. It is sown in May or June, transplanted at the onset of the rains and harvested on high ground between November and January, and on lower ground by February. The next important crop is the autumn crop, *Aus*, sown in April–July and harvested in August–December. Summer rice or *Boro*, sown in September–February and harvested in March–June, was until recently quite marginal, being mainly a marsh rice transplanted in winter. It has, however, been gaining momentum with the development of private shallow tube wells and the availability of high-yielding varieties. A study by Alok Bandyopadhyay and Boudhayan Chattopadhyay has shown how the owners of tube wells lease land from cultivators located within the command area of their wells who do not have enough resources to have their own tube wells or to buy water. This contract system at a fixed rate of produce per unit of land amounts to a seasonal reversal of the normal tenancy pattern.[7] In some cases, there are 'pure contractors' who either do not possess any land for cultivation or do not have their own land under summer cultivation. This tendency might find a limit in the indiscriminate exploitation of groundwater leading to a fall of water levels and ultimately to the drying up of aquifers.

South of the 17° latitude, rains start later. 'South of the latitude of

Nellore [around 14°] and excluding Kerala, while the date of onset of the South-West monsoon remains indistinct, it is possible to recognize very clearly a second "burst" in October at many rainfall stations, corresponding to the commencement of the retreating monsoon rains.'[8] Along the western coast of Kerala there is a clearly bimodal pattern of rainfall with a much longer rainfall period, nearly eight months in Calicut (Kozhikode), and becoming even longer further south along the coast. This local pattern allows three paddy crops in the coastal plains.

On the whole, progress in rice production has been disappointing. Although the area under high-yielding varieties of paddy has increased (according to official statistics) from 15% of the total 1970–1 rice area to about one-third ten years later, the recorded increase in yield remains smaller than anticipated, with an average of between 1000 and 1300 kg/ha. and an annual growth in yield of 1.14% for the period 1960–1 to 1976–7. These rather low average yields result from two main causes. The first is that 60% of the crop is grown under rain-fed monsoon conditions and is subject (often in the same year) both to moisture stress from dry spells and flooding from heavy rains. The second is that agricultural research has failed to provide the many varieties needed, adapted to these conditions in their local variations, and has tended instead to propose a few varieties over large tracts without paying enough attention to existing rainfall variability and micro-climatic properties. It should also be added that the traditional paddy areas often have a particularly large proportion of small tenants and sharecroppers and, therefore, a socio-economic structure which is not conducive to large investments in cultivation.

Irrigated wheat yield increases: how sustainable are they?

We have mentioned that the area under food grains is divided almost equally into three parts: paddy, wheat and pulses, and 'coarse' cereals. The areas under wheat and pulses are equal, around 22 million hectares each (respectively, 17.7% and 17.5% of the total area under food grains in 1980). We have combined pulses and wheat not simply to make a division into three parts easy to remember, but also because of the partial overlapping of their spatial distribution which is governed by climatic factors. From 1960–1 to 1976–7 the average annual growth in area cultivated was 3.37% for wheat (as against 1.14% for rice) and −0.37% for pulses. In fact, most (about 70%) of the expansion of the area under wheat was the result of increased cropping intensity – sowing of previously seasonal fallow land, which was facilitated particularly by the expansion of tube-well irrigation. But most of the remaining expansion of wheat area is accounted for by a decline in the area under pulses, by 22% according to a 1977 study by the International Crops Research Institute for the Semi-Arid Tropics (ICRISAT).[9] This study argues that the higher wheat yields more than compensated for their lower percentage protein content, as a hectare of wheat produces nearly 50% more protein than a hectare of pulses. This strange argument fails to take into consideration either the actual diet of the poor and the social desirability of pulses in the Indian diet, or the agronomic consequences of the increasing imbalance in crop rotation, and should be regarded as a measure of the bias in agricultural research and policy in favour of wheat. A measure of this neglect is the negative annual growth rate of pulse yields (−1.1%) for the period from 1967–8 to 1979–80.

Wheat is a typical example of a *rabi* crop, being sown in October–December during the last rainy days or when the soil still has a high moisture content, and harvested in April before the peak of the summer heat. The main wheat-producing areas lie in the states of Uttar Pradesh, Madhya Pradesh, Punjab, Haryana, Rajasthan, Maharashtra, Bihar and Gujarat. As wheat is grown in the *rabi* season, when there is very little rainfall in the major wheat-growing areas, the water retention capacity of the soil is an important factor. The bulk of the crop is grown on Indo-Gangetic alluvium soils, rather light in texture, with low organic matter content and, therefore, low water-retention capacity. Irrigation is thus particularly important.

Nearly 60% of the area under wheat is now irrigated as a result of the expansion, particularly of private tube wells, taking place in the late 1960s and early 1970s. In the period from 1960–1 to 1976–7, the average annual growth amounted to 3.37% in area and 4.12% in yield. These percentages are indicative of the effects of the so-called Green Revolution, which in India was mainly a 'wheat revolution'. This phenomenon has been the subject of many studies, including those carried out by UNRISD under the direction of Andrew Pearse, and so it is not necessary to dwell upon it here.

Let us only note that these so-called 'modern' methods of cultivation, very costly in terms of energy requirements (for irrigation pumps, and in the form of fertilizers and pesticides, for instance), have neglected the potential of traditional mixed-cropping methods. For instance, in poor fields in Madhya Pradesh, wheat is mixed with gram (chickpea).

The proportion of gram varies in accordance with the quality of the soil, a smaller quantity being sown in good soil and a large quantity in poor soil. In general, the average is about two-thirds wheat and one-third gram but in a year of short rainfall the proportion of gram is increased . . . In Rajasthan a popular practice is to grow a mixture of wheat and barley, the respective seeds being sown in alternate beds . . . pulses such as *Urad* (*Phaseolus* varieties), *Moong* (*Phaseolus mungo*), *Mot* (*Phaseolus acontifolius*) and cow peas (*Vigna catjang*) which are *kharif* pulses are sown mixed with *bajra*, *jowar* and maize . . . Peas are mixed with *rabi* crops like wheat and barley.[10]

This traditional intercropping of wheat operates at rather low levels of productivity, but protects against the risk of total crop failure were a single crop only to be produced. The potential intercropping will be dealt with below in relation to 'coarse' cereals. In the case of wheat, the rapid spread of a few high-yielding varieties has reduced heterogeneity, and, therefore, has increased the possibility of spreading the risks arising from climatic factors and from plant diseases and pests, which are themselves often affected by climatic factors. The 19th-century Irish famine was the most tragic example of the vulnerability of monospecific crops (particularly those which are homogeneous, as were potatoes in Ireland, being derived from a few clones): the potato harvest was devastated by a single pest (*Phytophthora infestans*), the development of which was favoured by the weather conditions of 1846.

The introduction of highly homogeneous varieties of wheat in northern India has increased the demand for labour at harvest time, which occurs on the same dates over large areas. Hired agricultural labourers are, therefore, in a stronger position to exert pressure on large farmers for increased wages. Faced with such a social 'threat', wealthy farmers tend to mechanize their harvest operations with combine-harvesters, thus skipping the 'reaper' phase of the development of mechanization in Europe and Northern America.[11] This had two effects: labour requirements for threshing were reduced, and the immediate release of grain for the market became possible. Such cases of mechanization are still few in number but are illustrative of the 'modernization' of wheat farming and, at the same time, indicate its spatial and socio-economic limits. In addition, soil humus content seems to deteriorate, lowering the water-retention capacity of alluvium.[12] More irrigation thus becomes necessary, increasing the demand on edaphic reserves, already depleted by the wild proliferation of deep tube wells which are bringing down the water table. Agronomic and hydrologic factors, as much as socio-economic ones, are constraining further wheat yield increase if not contributing to potential stagnation and even decline in the future.

Coarse cereals: the neglected crops of the poor

Another third of the area under food grains is devoted to coarse cereals, with jowar (*Sorghum vulgare*) making up 10% of the total food grain area, and bajra (*Pennisetum typhoideum*) only 7%. Jowar and bajra are the staple food of the poor in large areas of the country, particularly of the Deccan. More than 95% of the area under these crops is rain-fed, so their yields fluctuate widely due to weather variations. Despite some rise in jowar yields (1.14% for the period from 1960–1 to 76–7), total production has increased very little (0.1%); the area under jowar has shrunk (−1.04%) as, with the spread of irrigation, more profitable crops have taken its place. Average jowar yield is still low, under 600 kg/ha. It is, however, the third most important cereal in India, after rice and wheat. Maharashtra is the leading producer state with an output of about 4 million tons. Andhra Pradesh, Karnataka and Madhya Pradesh are the other major producing states, with an output of more than 1 million tons each. Practically no jowar is grown in eastern India. Jowar (known as 'cholam' in the south) is both a *kharif* and a *rabi* crop; as *kharif*, it is mainly grown on black soils, often rotated with cotton and sometimes still mixed with pulses or sesame; *rabi* jowar, less often mixed with other crops, needs moisture-retaining soils which have received plentiful rain in September–October, and is perhaps most notable in Tamil Nadu. Bajra (or 'bajri'; 'cumba' or 'cambu' in the south) tolerates lighter soils than jowar, and is, therefore, grown extensively on the poorer Deccan lava uplands and on sandy or stony soils generally in western and southern parts of India. It is not grown in eastern India. By and large, it is a *kharif* crop, but is harmed by too much rain and so is usually sown after the first force of the rains is spent. Bajra is usually mixed with pulses, so that rotation is not really necessary, but it may be 'rotated' with jowar in weak monsoon years.[13] In terms of area, it is the fourth most important cereal crop in India, but because of its low yield, ranks fifth in terms of output.

Ragi ('marua', 'mandua', or 'mandia'; *Eleusine coracana*) ranges from the Himalayan slopes almost to the extreme south. Again, it is a *kharif* crop, which may be transplanted. It is rarely intercropped, except in tribal areas of Orissa where shifting cultivation is practised. Maize is mostly cultivated north of the 16° latitude, and in terms of total output is the fourth most important crop, with Uttar Pradesh, Punjab, Rajasthan, Bihar and Madhya Pradesh accounting for more than half the total maize output.

Sorghum and millet are typically crops of the rain-fed semi-arid tropics (SAT) of India, for which it is absolutely essential to relate rainfall to the water storage capacity of the different soils of the region. Two major soil types are found in the area of extension of sorghum and millet: alfisols (red soils), which are light and drought-prone, and vertisols (black soils), which have great water-retention capacity. It has been observed that about 18 million hectares of deep vertisols in India are being fallowed during the rainy season. The low productivity of post-rainy season (*rabi*) crops grown on residual moisture seems to indicate inefficient use of available water. The exposure of uncropped soils to intense rains appears to result in serious runoff and erosion in spite of the presence of soil conservation structures.[14]

N. S. Jodha[15] lists the reasons advanced by farmers for fallowing the deep vertisols during the rainy season:

1.   In the absence of good soaking rains, the deep Vertisols are too hard to work; once substantial rains begin, it is difficult to enter such fields.
2.   Even if some crops are dry sown in deep Vertisols prior to rains, the management of the crop during the subsequent wet period is difficult. Weeds may ruin the crop before the fields are dry enough to permit entry of labor.
3.   The rains received during the early phase of the monsoon are less dependable than those received during the later phase. According to the farmers' experience and meteorological data, early rains are inadequate to fully saturate the profile of deep Vertisols. The crops planted during the first phase of the monsoon are exposed to the risk of drought in a prolonged midseason dry spell, and to waterlogging as well as increased disease incidence caused by continuous rains in the second phase of the monsoon when they are at the flowering or ripening stages.

At present, farmers – not aware of crop varieties or land management practices that can reduce these hazards of rainy-season cropping in the deep Vertisols – continue to follow the traditional practice of fallowing land in the monsoon season. Given the hazards of rainy-season cropping and the nonavailability of viable technology to counter them, the farmer probably makes a rational choice in leaving the deep Vertisols fallow during the monsoon. The irrationality of rainy-season fallow can be demonstrated only by providing a viable alternative, and this constitutes the challenge for agricultural research.

Even if one ignores the benefits of reduced soil erosion when Vertisols are planted in the rainy season, the potential payoff from a breakthrough in technology for monsoon-fallow areas, facilitating the raising of *kharif* crops as well as *rabi* crops, will increase the gross cropped area by nearly one-fourth of the current net sown area in SAT India.

Furthermore, since small farmers leave a higher proportion of their land fallowed during the monsoon than do large farmers, the prospective low-cost technology for such areas may help small farmers more than large ones.

Table 3.1 Percentage share of different crops in the gross irrigated area in six SAT Indian villages (average of 1975–6 and 1976–7).[a]

	Proportion of different crops in gross irrigated area[b]					
Crops	Kanzara	Kinkheda	Kalman	Shirapur	Aurepalle	Dokur
Sorghum	28[c] (26)	–	30 (28)	9 (3)	6 (6)	–
Wheat	56[d] (58)	44[d] (45)	19 (23)	15 (14)	3 (2)	–
Paddy	2 (2)	1 (1)	6 (5)	1 (1)	73 (78[e])	79[e] (74)
Groundnuts	6 (5)	10 (9)	4 (4)	10 (10)	–	20 (24)
Pulses[f]	5 (4)	25 (27)	9 (5)	2 (1)	–	–
Vegetables	4 (4)	2 (1)	7 (11)	12 (13)	5 (6)	1 (1)
Sugar cane	–	–	3 (6)	22 (39)	–	–
Cotton/castor bean[g]	–	9 (13)	–	–	5 (2)	–
Other sole crops[h]	–	–	13 (13)	23 (17)	–	1 (1)
All mixed crops[i]	–	9 (4)	10 (7)	6 (3)	10 (6)	–
Total	100 (100)	100 (100)	100 (100)	100 (100)	100 (100)	100 (100)
Total weighted irrigated area as % of gross cropped area	5 (22)	3 (6)	8 (21)	13 (42)	19 (66)	60 (307)

[a] Based on details from 180 sample farms in six villages. Village-level studies have been conducted in these villages since May 1975 (Jodha *et al.* 1977). The sources of irrigation are tanks and wells in Mahbubnagar and Aurepalle and wells in other villages.

[b] Figures in parentheses indicate the proportion of each crop in the gross irrigated area recalculated using the intensity of irrigation. The recalculated weighted gross irrigated area is based on area irrigated multiplied by number of irrigations given to the same (whole) plot. All irrigation operations for a given plot taking place within a ten-day period have been treated as one irrigation operation to avoid the possibility of partial coverage of a plot by water being treated as its full coverage. Partial coverage may result from poor and slow recharge in the irrigation well and the water-spreading methods used in paddy fields. In the case of paddy, this method tends to underestimate the irrigation intensity, because watering of paddy is almost continuous and the field is always kept wet.

[c] Hybrid sorghum.

[d] Kanzara over 60% and Kinkheda over 60% high yielding variety (HYV) wheat.

[e] HYV paddy over 60% and 90% respectively in Aurepalle and Dokur.

[f] Mung bean in Kanzara and Kinkheda; chickpea in Kalman and Shirapur.

[g] Hybrid cotton in Kinkheda; castor bean in Aurepalle.

[h] Includes maize, sunflower, garden crops in Kalman and Shirapur villages, and finger millet in Dokur.

[i] Excludes all vegetables, mixtures and a limited extent of sugar cane–vegetable mixtures, included with respective main crops.

Allowing the soil to lie fallow during the rainy season is not absurd. It is a rational choice of farmers, particularly small farmers, given the technology available to them, and it offers a challenge to agricultural research. In the rain-fed regions, where sorghum and millet are grown and which do not have deep vertisols, another agricultural practice is dominant and generally considered as a 'backward' practice: intercropping. According to Jodha, it 'covers 35 to 73% of their gross cropped areas. The extent of intercropping declines with increases in irrigation in villages. The small farmers again have a significantly higher extent of intercropping than large farmers. This indicates that generation of a low-cost new technology for intercropping may help less endowed areas and farmers more than the relatively well-endowed ones.'[16]

Jodha reports that in the six villages selected by ICRISAT,[17] two-crop mixtures were popular in most villages (period 1975–6 to 1977–8), but mixtures involving five to eight crops were not uncommon. Table 3.1 gives the proportions of different categories of crop mixtures over the total area of intercropping in these six SAT villages. Research is in progress to analyse rainfall and cropping decisions using village-level daily rainfall records and cropping patterns on sample farms. Jodha summarizes the practice of the SAT villages in the following manner:

> As revealed by the number of crop combinations (as high as 84 in a single village), traditional intercropping is highly complex. This is partly an outcome of farmers' informal experimentation with crops that could satisfy their requirements and also fit the agricultural environment of the region. The multiple objectives of the farmer such as security, profitability, employment and subsistence requirements of his family members and cattle etc., should be taken into account when evolving new intercropping technology.
>
> The juxtaposition of requirements of prospective watershed-based technology and the features of the traditional system of farming – particularly the land ownership and usage pattern – gives an idea of the institutional constraints the technology is likely to face. Because integrated watershed-based technology is indivisible and because individually-owned land parcels large enough to constitute a composite mini-watershed are not available, there seems no alternative to group action that can ensure management of land for higher productivity and conservation on a watershed basis. In order to induce group action among farmers, prospective watershed technology will have to be highly profitable.[18]

The superiority of intercropping, in terms of higher gross returns as well as higher and more evenly spread employment of labour when compared to single cropping, has been documented for many years by isolated researchers,[19] who were not taken seriously by the scientific establishment until the past few years, when they have been given more legitimacy.

Agricultural research: for whom?

To demonstrate, at the technical level, the potential of intercropping systems for a more productive and more resilient agriculture also entails, at the ideological level, a break with the dominant paradigm embodied in the so-called 'Green Revolution', which counterposes the 'modern' farmers of the Punjab using fertilizers, pesticides and machinery on irrigated wheat as the sole crop, to the mass of 'backward' peasants cultivating coarse cereals in rain-fed areas and letting their land lie fallow during the rainy season, or practising intercropping. At the international level, the predominant trend in agriculture has so far been more towards universal solutions than specific ones.

Such endeavours, supported financially and intellectually by the major donor countries, are not without consequence for the orientation of national agricultural research in developing countries. In India, since the late 1960s, major efforts have been developed to foster such 'modern agriculture' based on homogeneity (one crop only in one field, and one variety only, preferably a superior cereal, mostly wheat); plant breeding for maximum yields (with not enough attention paid to short-term variability and long-term sustainability); high energy costs derived from the use of irrigation water, fertilizers, pesticides and machinery, all inputs requiring an efficient support system (pumps, machine spare parts, fuel or electricity, and so on) for timely delivery; and price subsidies.

This approach derives from the agricultural model of the industrialized countries, which has been shaped by completely different circumstances. It well fits the conditions of production of a small minority of farmers in India, spatially and socially well delimited, but cannot be extended to all regions and to all socio-economic groups, because of its energy costs and its general consequences, especially for employment.

At the international level, a new trend has been perceptible, however, since the establishment of ICRISAT in India in 1972. Less well-endowed than its counterparts dealing with maize and wheat (International Centre for the Improvement of Maize and Wheat, CIMMYT, Mexico) or rice (International Rice Research Institute, Philippines), it is not organized around one crop but deals with a zone defined by climatic factors. The need to study farming systems is gaining ground as it was clear from the beginning that a holistic, interdisciplinary approach to research on soil, water and crop management and the integrated application of new technology would be essential to successful agricultural development in the Semi Arid Tropics (SAT). Single component approaches were not expected to solve the complex problems encountered in the SAT.

Undependability of the crop-water environment is the most limiting factor to crop production; therefore earlier approaches consisting of only variety improvement, crop and fertility management, fallowing, or bunding, etc. could not result in substantial effects. Emergency relief remained common in many areas. Even conventional irrigation projects were considered inadequate to meet the real needs of agriculture in the SAT.[20]

This statement renews the approach of Sir Albert Howard, who spent his life in India studying farming practices and directing agricultural research before World War II; his 'agricultural testament' with its insistence on maintaining soil fertility, using organic manure, and so on, was ignored or laughed at by younger generations more interested in the prospect of patterning Indian peasants after their Minnesota or Iowa farmer colleagues. For Sir Albert Howard, 'Nature' always raises mixed crops:

> Instead of breaking up the subject into fragments and studying agriculture in piecemeal fashion by the analytical method of science, appropriate only to the discovery of new facts, we must adopt a synthetic approach and look at the wheel of life as one great subject and not as if it were a patchwork of unrelated things.[21]

Sophisticated agricultural research is necessary to understand intercropping practices and improve on them: each species has its own requirements for water and nutrients, according to a specific time pattern, its own root system exploring different layers of the soil, its own physiology (nitrogen–fixation), and so on.

The identification of areas of complementarity or synergy between two species[22] in order to make full use of soil resources, temperature and rainfall regimes requires a tremendous scientific effort which should contribute to the improvement from within of the local practices by understanding them better, and to increasing productivity in a sustainable way, without high energy costs and long-term degradation of soils. In areas of high rainfall variability such research could contribute to the development of cropping systems less vulnerable to climate variability and requiring a better seasonal distribution of labour. But research can be successful only if it develops two-way channels between scientists and the poorer agriculturalists – those whose food security should be improved – and does not confine itself, as is the case at present, to the better-endowed farmers. Agricultural research and extension should reply to the unverbalized needs of the majority and not only to the effective social demand of the few.

In order to do that, farming systems must be analysed in the total socio-economic framework in which they are set. The holistic approach as well as the small-farmer focus are far from being accepted. An example of the reluctance of international circles to direct research to the needs of the majority is given by Edmund K. Oasa in his contribution to this volume (Ch. 2 above). He refers to the Technical Advisory Committee (TAC) quinquennial review of ICRISAT:

> The review team questioned this institute's targeting of the small farmer and the consequent emphasis on zero or low-input farm conditions. The team, which conducted its review in 1978, argued that ICRISAT might be 'criticized for perpetuating agricultural under-development'.[23]

However, L. D. Swindale, the institute's Director General, maintained a year later that

> The principal target of ICRISAT's research is the small farmer of limited means, who often has poorest access to technology and the markets that can help him effectively benefit from it.[24]

Seasonal rural unemployment: a critical issue for the majority

One might question the TAC team's obsession with annual production figures at the national level, ignoring the hunger months of small farmers and agricultural labourers. It was recently officially estimated that about 11% of the annual increase in the labour force is absorbed in the organized sector outside agriculture.

According to the Government's Rural Labour Enquiries, there was a 78.6% increase in the number of agricultural labourer households between 1964–5 and 1974–5, i.e. from 5.7 million to 10.2 million, compared to an increase of only 16.6% in the total number of rural households.

With an agricultural labour force representing more than 70% of the total economically active population, with such a growing reserve army of labour depending mainly on agriculture, and with employment opportunities outside agriculture so restricted, the seasonality of employment and income is a major feature characterizing the Indian food system. According to the 1976–7 Agricultural Census, nearly three-fourths of the 81.5 million operational holdings are below two hectares, with 55% of all holdings being less than one hectare (against 51% in 1970–1) and representing 10.7% of the total area operated. These families cultivating less than two hectares in India, often as tenants or share-croppers, number around 60 million and represent a population of more than 300 million. The extent to which these families are able to earn some supplementary income during the off-season critically affects their food intake. In order to obtain enough food, many of them need to take advantage of the peak demand for labour of larger farmers even when, due to climatic constraints, this sale of labour conflicts with their own labour requirements. The same labour time, therefore, has different values on the calendar of seasons, only partially reflected in the seasonal wage rate fluctuation.

At the other end of the spectrum, we find holdings of ten hectares and more representing 3% of all holdings in 1976–7 and accounting for 26.5% of total operated area. Most of the farmers hire agricultural labourers for their agricultural operations for a total number of days, often of the order of half a year. These agricultural labourers, helped by casual labourers during peak periods, must find resources to tide them over their periods of unemployment, whether days, weeks or months. Whereas economic analysis of the farms employing agricultural labourers is relatively easier (although we have mentioned that agricultural labourers might be employed for non-economic reasons such as a caste restriction on manual labour), such analysis is much more difficult for the intermediate group of cultivators who usually neither work for others, nor hire labour, except marginally. Such a situation means that they can largely accommodate the peak periods with their own family labour (sometimes supplemented by that of relatives on a free exchange basis). If labour time is not a constraint, and if all income derives from agricultural pursuits, value becomes more elusive. To work ten days less for the same output means to save the extra energy needed for those ten days, but does not change the overall productivity of labour which must be assessed within the framework of the totality of the agricultural year: all family members, working or not, have to live throughout the year and it is the cost of reproduction of the labour force over the year which must be taken into consideration.

A capitalist farmer employing an agricultural labourer for peak periods is not concerned about what happens to this labourer's family at other times of the year. The Jajmani system[25] should be viewed in the context of a certain scarcity of labour and of local social relations. When there is a dissolution of these pre-capitalist forms, the only limits to the unconcern of large farmers for the starvation of agricultural labourers during the off-season are set by the necessity to maintain some social order – consumption loans given to agricultural labourers buy social peace while forcing wages down further. When some forms of organization of agricultural labourers takes place, as, for instance, in the Punjab at harvest time, large farmers are prompted to mechanize harvesting operations: to buy (or rent) a combine-harvester might not be economically sound but becomes a social necessity. The building of an over-capacity in agricultural machinery to surmount the shortages of peak periods has a long history in the market economy of industrial countries. Indeed, it still has a future for some sectors of production (e.g. fruits, vegetables) even as the alternative solution of drawing upon seasonal migratory workers persists – workers who are all the more vulnerable when they come from far away and particularly docile if, in addition, they are foreigners and most often illegal immigrants. Seasonality of production, and peak periods of labour should be perceived, in greater measure than is the case at present, as a major dimension in the understanding of modern capitalist agriculture.

In India, the development of pockets of capitalist agriculture has triggered the seasonal migration from different regions of groups of agricultural labourers competing for employment. The distance travelled is an indication of the destitution of these workers in their own villages. It was found in the course of the field survey of the Food System and Society project in Eastern India that of about 250 households in a Sundarban area of South 24-Parganas (West Bengal), all adult male members of as many as 60 households migrate every year as farm labourers to the Nainital Terai in Uttar Pradesh, 1400 kilometres away.

Although interesting results have been achieved locally in activating or reactivating handicrafts and cottage industries in India, the ancient socio-economic fabric of pre-British India cannot be restored in the international context of today, and no large-scale solution to seasonal hunger and deprivation can be achieved in that manner. Agriculture will remain for a long time to come the main source of employment and income. As, in India, the limiting factor and the leading input in the development of a year-round agricultural activity is water, considerable efforts have been made by the Government to develop irrigation schemes, but with a certain bias in favour of large-scale irrigation projects. At the same time, dry-farming techniques have not been developed to the desirable extent. Many dry-farming programmes are, in fact, not geared to dry farming but to irrigation, with the considerable water losses that canal irrigation entails in arid zones.

But ground water has been tapped on an increasing scale since the early seventies, mainly by private farmers, in an unplanned manner which threatens its very existence. According to P. R. Pisharoty and P. Sharma of the Ahmedabad Physical Research Laboratory:

A survey in the Rajasthan desert area conducted by the Central Arid Zone Research Institute (CAZRI), Jodhpur (Chatterji, 1978) has revealed that the utilisation of ground water is increasing at a very fast rate, especially in the last one decade. In some districts of Rajasthan the annual utilisation of ground water, in fact, far exceeds the recharge. Nearly 65 per cent of the total area of Western Rajasthan has saline ground water which cannot be profitably used, unless proper management practices are adopted.

The ground water in the deep aquifer was perhaps stored a few thousand years ago. Recent age determination of the ground water at a few places in the Sabarmati basin, using 14_C method, indicated the apparent age to be between 2000 years close to recharge areas and 20,000 years near the discharge areas of confined aquifers (S. K. Gupta, private communication). In view of the small ground water recharge area (of the order of 5–10%) and the present trend of over-exploitation of ground water, it seems unlikely that groundwater will be a long term solution to the water problem of the drought prone areas unless large scale artificial recharging of aquifers is resorted to.[26]

Further studies in geology and hydrology are, therefore, necessary in order to plan, in a rational way, sustainable in the long term the coordinated use of ground and surface water.

Conclusion

As already emphasised, the main efforts of agricultural research and extension services have concentrated on the so-called 'superior cereals', and particularly on irrigated wheat.[27] But in order to achieve year-round cultivation, a major campaign must be mounted in relation to a variety of other food-crops, to relay-cropping and intercropping. Much has to be learned from the farmers themselves concerning these complex crop systems, if agricultural scientists are willing to listen not only to the more affluent in the irrigated areas, but also to the poorer in rain-fed areas.

In order to provide adequate food self-provisioning and/or food purchasing power to almost three-quarters of India's 700 million inhabitants whose livelihood depends on agriculture, and sufficient food supplies to the remainder, complex agricultural systems must be developed through the interaction of scientific and vernacular knowledge. This implies that the quest should not be for solutions in terms of large regional units, but for local solutions in each ecosystem and that climate especially should be treated more as a resource than a constraint, in order to fully utilize the human and natural resources of each microsystem.

At the national level, to devote a large share of overall agricultural research funding to coarse cereals,[28] rain-fed agriculture and small farmers, calls for hard political decisions, which presuppose the existence of pressures organized by small farmers.

This requires the existence of institutional means to enable them to organize themselves, not only around technological problems, but also around land tenure and credit issues. It is in such a socio-economic framework that agricultural research should be organized in a more participatory way and in a more interdisciplinary manner.[29] It should be future oriented, so as to guarantee the conservation of the natural resources (soil, water, genetic resources) on which agriculture itself is based.

To divorce technological issues from institutional ones, to counterpose them and to submit that a different technology alone will eliminate food insecurity, is not only to reproduce, at a different level, the Green Revolution paradigm, but is also to forget that, contrary to the Green Revolution, there are at the present time no social forces sufficiently well-organized to provide such endeavours with the massive support required.

This does not mean that modifications are not possible. But they require ideological changes in the system of education in general, and of agricultural education, research and extension in particular, towards a more humble approach to the complexities of the biological and social systems, with the working peasant at the centre of the stage.

Notes

1 US Government Printing Office (1970).
2 Ladejinsky (1969a).
3 Ladejinsky (1969b).
4 Publications available at UNRISD, Palais des Nations, 1211 Geneva 10.
5 Spitz (1983).
6 1978–9 data from Ministry of Agriculture (1980).
7 Bandyopadhyay and Chattopadhyay (1979).
8 Ramamurthy (1972), p. 29.
9 Ryan and Asokan (1977).
10 Indian Council of Agricultural Research (ICAR) (1968), pp. 50, 98–100.
11 During the reaper phase, which was still prevalent in Europe in the decade following World War II, farm-by-farm threshing was a process which took several months to complete. Grain was released only gradually to the market, thus reducing sharp price in hectares after the harvest.
12 The general soil fertility maintained in the Punjab for generations by careful soil management practices has been eroded in a few years of

intensive use of fertilizers. Wolf Ladejinsky pointed out as early as 1976 that soil tests have shown that soils are being depleted of phosphorus, which, not needed ten years ago, has become essential, as well as micro-nutrients such as zinc, sulphur and iron (Ladejinsky 1976, p. 4).

13 See Spate (1954), pp. 215–18.
14 Kampen (1979), p. 42.
15 Jodha (1979), pp. 13–14.
16 Ibid.
17 International Crops Research Institute for the Semi-Arid Tropics, Patancheru, Andhra Pradesh, India.
18 Jodha (1979), p. 23.
19 See, for instance, Mathur (1963) and D. W. Norman, work in Nigeria, quoted by Jodha (1979).
20 Kampen (1979), p. 42.
21 Howard (1940).
22 Or even two varieties, as Professor Subodh Kumar Roy of the Indian Statistical Institute, Calcutta, has been doing for many years in relation to paddy, with very little financial support. It is only recently that the Indian Council of Agricultural Research has demonstrated more interest in such a research perspective.
23 TAC (1978), pp. 13–14.
24 In Ryan and Thompson (1979), p. vii.
25 A traditional system, often described in the anthropological literature, through which village officials, service and artisan castes were receiving their remuneration in kind at harvest time, as a share of the cultivator's grain heaps.
26 P. R. Pisharoty and P. Sharma, 'Report on Drought and Man' for the IFIAS 'Drought and Man' project.
27 The Pantnagar Agricultural University in Uttar Pradesh is situated in rich, fertile land reclaimed from a forested area after World War II. It derives part of its income from a seed farm of several thousand acres, which is a model of scientific agricultural management. It is worth noting that the demand of agricultural workers for year-round employment and monthly wages triggered campus violence. If such a farm is not able to provide an adequate living wage for its agricultural workers throughout the year, it is difficult to admit that such models should be replicated all over the country.
28 The same could be said about shifting cultivation practices in tribal areas, and also about the necessary adaptation to micro-ecological regions of rice varieties and local paddy cultivation practices.
29 See P. Spitz (1935).

References

Bandyopadhyay, A. and B. Chattopadhyay 1979. Contract system of summer rice cultivation. A Bengal case study. *Business Standard* (Calcutta), 11, 12 and 13 October. Now available in a longer version in *Cressida Trans* **1**(1).

Bhalla, G. S. and Y. K. Alagh 1979. *Performance of Indian agriculture*. GOI Planning Commission. New Delhi: Stirling Publishers.

Howard, A. 1940. *An agricultural testament*. London: Oxford University Press.

Indian Council of Agricultural Research 1968. *Farmers of India*, vol. IV. New Delhi: ICAR.

Jodha, N. S. 1979. Some dimensions of traditional farming systems in semi-arid tropical India. In *International workshop on socio-economic constraints to development of semi-arid tropical agriculture*. Patancheru: ICRISAT.

Jodha, N. S., M. Asokan and J. G. Ryan 1977. *Village study methodology and resource endowments of the selected villages*. ICRISAT Economics Program occasional paper 16.

Kampen, J. 1979. Farming systems research and technology for the semi-arid tropics. In *Development and transfer of technology for rainfed agriculture and the SAT farmer*. Patancheru: ICRISAT.

Ladejinsky, W. 1969a. The Green Revolution in Punjab. *Econ. & Pol. Weekly* (Bombay) **4**, 26 (June), 73–82.

Ladejinsky, W. 1969b. The Green Revolution in Bihar: the Kosi area. *Econ. & Pol. Weekly* (Bombay) **4**, 39 (September), A147–A162.

Ladejinsky, W. 1970. Ironies of India's Green Revolution. *Foreign Affairs* (July).

Ladejinsky, W. 1976. Agricultural production and constraints. *World Development* **4**(1).

Mathur, P. N. 1963. Cropping pattern and employment in Vidharba. *Indian J. Agric. Econ.* **18**, 38–43.

Ministry of Agriculture 1980. *Bull. on food statistics*. New Delhi.

Ramamurthy, K. 1972. *A study of rainfall regions in India*. University of Madras.

Ryan, S. G. and M. Asokan 1977. *Effect of Green Revolution in wheat on reduction of pulses and nutrients in India*. ICRISAT, Occasional Paper 18.

Ryan, J. G. and H. L. Thompson (eds) 1979. *Socioeconomic constraints to development of semi-arid tropical agriculture*. Patancheru: ICRISAT.

Spate, O. H. K. 1954. *India and Pakistan*. London: Methuen.

Spitz, P. 1983. *Food systems and society in India – a draft interim report*. 2 vols. Geneva: UNRISD.

Spitz, P. 1985. Food systems and society in India: the origins of an interdisciplinary research. *International Social Science Journal* **XXXVII**(3), 371–88.

Technical Advisory Committee (TAC) 1978. *Farming systems research at the international agricultural research centers*. Washington, DC: CGIAR.

US Government Printing Office 1970. *Symposium on science and foreign policy: the Green Revolution*. Proceedings before the Subcommittee on National Security Policy and Scientific Development of the Committee on Foreign Affairs, House of Representatives. Ninety-first Congress, First Session, 5 December 1969.

PART II

Alternative approaches in three continents

4 Alternative developments in Brazil

ADEMAR RIBEIRO ROMEIRO

Alternative developments required in Brazilian agriculture to produce enough food, not only for export but also for sustaining the population, must take into account both the ecology and the human resources available in the country. But first we must consider the historical events which have resulted in food scarcity, the predatory exploitation of nature and chronic unemployment existing side by side with the commercial success of export crops. What has caused food scarcity and rural unemployment in a country with an agricultural area, in 1975, of 323 million hectares (three times the agricultural area of China or 2.5 times that of India), and an estimated area of 100–200 million hectares yet to be exploited in the Amazon region (FIBGE 1981)? Obviously, an exhaustive study of over four centuries of agricultural development cannot be made in a short chapter, but we can point out some aspects which we judge necessary to understand the problems brought about by the recent Brazilian agricultural expansion.

Historical perspective

The colonial period

Brazil's basis for agricultural development was the large plantation producing for export. Food production for the internal market has always been considered secondary. Some took place on the poorest lands of the large farms to provide food for the workers (or slaves); some was undertaken by free workers living on lands lying between the *latifundios* (where an important labour reserve for the large landlords was concentrated, a mass of population obliged to supplement their income by working for the landlords, because of the insufficient plot of land (*minifundio*) allowed them); finally, food production took place on the ever-expanding agricultural frontier. Meat production was carried out by the large farms involved in extensive cattle breeding. This kind of farm played an important role in land concentration and monopoly.

It is important to note that food production occurs in the areas residual to export agriculture (in the interior or on the outskirts of the large properties) or in areas of no interest to the powerful owners (the agricultural frontier). The result of a precarious and temporary use and possession of land for food production was its instability and the chronic supply problems that have been observed since the 17th century. These problems were a constant preoccupation of the Portuguese kingdom, which tried to provide for the supply of cities and villages through legal measures: *Provisão* of 24 April 1642 made it obligatory to grow cassava (the 'bread of the land') on an area equivalent to that of the export cultures; *Alvara* of 25 February 1688 compelled the inhabitants of Bahia to grow 500 *covas* of cassava for each slave owned; other legal measures of this sort followed (like the *Carta Regia* of 11 January 1701), but the landlords always resisted their application (Linhares & Teixeira da Silva 1981). The chronic food scarcity together with a disorganized commercial structure resulted in the formation of a group of 'intermediaries' which became very strong and which throughout Brazil's history has kept food prices low for the farmer and high for the consumer. The need to combat these groups of *atravessadores* (the pejorative name for dishonest commercial intermediaries) became another constant preoccupation, at least in the political rhetoric.

With the decline of slavery (formally abolished in 1888) the process of land monopolization by the rural oligarchy, whose farm states were known as *latifundio*, gained impetus. Wage workers were needed to replace the slaves in the coffee plantations, thus it became necessary to prevent by any means the establishment by the European immigrants of an autonomous peasantry on the unused land. In this context the Land Law of 1850 appeared, providing the juridical instrument necessary to force the free peasant to become a wage labourer. This law, inspired by E. G. Wakefield's theories of colonization, determines that all government-owned, unused land can only be acquired through money payment, making it inaccessible to the landless workers. So the landownership structure continued to be dominated by the large plantations and cattle-breeding farms, except in certain limited areas (especially in the south of the country). And food production for the internal market continued to come from the poorest land, the agricultural frontier and, in the case of the coffee regions, between the rows of coffee plants during the first growing years.

Table 4.1 Landownership structure in Brazil, 1920–70.

Number of properties (percentage)	Area percentage				
	1920	1940	1950	1960	1970
50% of the smallest properties	3.9	3.6	3.2	3.1	2.9
10% of the largest properties	76.0	76.7	78.3	78.0	77.6
5% of the largest properties	66.1	66.2	67.9	67.9	66.8
1% of the largest properties	41.9	43.0	44.6	44.5	42.8

Source: Modified from Sorj, B. 1980.

The first phase of the industrialization process

By the end of the 1920s, because of the crisis in the export sector and the take off of the industrialization process, the agricultural sector had undergone an important change in its production structure, with the partial conversion of plantations to food and raw materials production for the rapidly expanding urban and industrial market. But the landownership structure remained extremely concentrated. As Table 4.1 shows, 1% of the largest properties covered more than 40% of the total agricultural area from 1920 to 1970, while 50% of the smallest properties covered only 3% of the total area in 1920, falling steadily to 2.9% in 1970. Just 5% of the largest properties covered almost 70% of the agricultural area during the whole period under consideration. The other main characteristics of the landownership structure, during a period of rapid industrial growth and urbanization followed by high inflation rates and other imbalances, started a great debate (see Carvalho 1978) in the late 1950s and early 1960s about the role of the agricultural sector in these problems. The question raised was whether an extremely concentrated landownership structure, based on the *latifundio*, would allow agriculture to play the role traditionally attributed to it in the economic development process.[1]

Tables 4.2 and 4.3 show clearly the dynamic functioning characteristics of Brazilian agriculture in this period. First, we note that expansion in production is mainly the result of growth in the cultivated area. Land productivity varies widely according to time period, region and crop, but for the whole period it is stagnant. This shows the low level of the agricultural techniques in use. Almost all crops, at different times, show negative rates of productivity growth that result from the predatory character of the techniques in use. A very good example is coffee cultivation. It moves from Rio de Janeiro state to São Paulo state through Paraíba's valley, leaving a

Table 4.2 Cultivated area and land productivity growth rates of the principal food crops, 1948–63 (percentages)

Period	North				North-East				South-East[a]				South				Central-West				São Paulo			
	Rice	Beans	Corn	Cassava	Rice	Beans	Corn	Cassava	Rice	Beans	Corn	Cassava	Rice	Beans	Corn	Cassava	Rice	Beans	Corn	Cassava	Rice	Beans	Corn	Cassava
1948–53																								
area	3.7	17.2	5.8	3.2	11.8	5.2	3.7	2.6	3.5	1.1	2.2	1.0	5.9	5.7	5.1	4.6	8.9	7.4	8.7	4.2	1.4	0.8	1.8	-0.9
productivity	-0.5	-0.5	-0.6	-1.7	-1.1	-3.8	-4.3	-2.2	2.1	0.3	-1.7	0.3	0.1	0.2	-0.4	0.6	-1.2	1.8	2.7	1.6	-1.6	1.4	-1.5	0.2
1953–8																								
area	8.4	-6.6	5.9	3.1	6.3	2.6	2.2	2.1	0.7	2.7	3.0	3.1	6.1	2.6	3.3	2.7	14.2	6.7	8.6	8.0	-0.7	0.5	0.8	5.9
productivity	-2.5	1.3	-4.3	0.3	0.8	1.8	1.7	-0.2	3.9	0.6	1.3	0.2	-2.1	-1.1	0.9	-0.4	-0.5	-0.6	-1.6	0.4	2.2	-0.7	2.6	1.1
1958–63																								
area	5.7	6.6	5.8	8.4	10.6	8.9	8.1	4.5	5.3	-0.4	2.7	2.7	6.0	6.7	4.9	6.5	16.1	9.9	13.0	8.7	2.5	-0.06	3.8	12.5
productivity	-0.4	0.2	1.6	3.3	2.6	0.1	1.7	1.5	-0.9	-4.0	-0.8	0.04	2.6	0.8	0.1	1.7	-1.1	-3.4	0.7	-0.1	-2.4	-4.3	-0.2	0.2

Source: Alves 1981.
[a] Not including São Paulo.

Table 4.3 Cultivated area and land productivity growth rates of the principal export crops, 1948–63 (percentages).

Period	North-East			South-East[a]			South			São Paulo		
	Sugar	Cotton	Coffee	Sugar	Cotton	Coffee	Sugar	Cotton	Coffee	Sugar	Cotton	Coffee
1948–53												
area	4.2	3.1	–	1.0	12.6	1.3	1.4	10.9	12.8	10.1	0.3	0.08
productivity	–0.5	–2.8	–	0.4	–1.4	–1.5	4.7	–4.0	–1.9	0.3	3.8	–1.0
1953–8												
area	3.7	5.0	–	3.9	6.2	3.7	1.8	11.0	22.6	8.6	–9.3	2.2
productivity	0.4	0.3	–	–1.8	–1.7	–6.2	1.8	6.5	–0.5	2.1	3.4	3.6
1958–63												
area	4.3	6.6	–	0.8	4.7	0.2	3.3	13.4	7.6	5.9	–0.3	–4.9
productivity	1.0	2.4	–	0.2	–1.1	4.4	2.6	4.0	–0.5	0.4	5.6	3.6

Source: Alves 1981.
[a] Not including São Paulo.

trail of depleted land. At the beginning of the period (1948), coffee reached the northern part of Parana state (southern region in Table 4.3), once a region of extremely fertile virgin land. As we can see, from the beginning the rapid expansion of cultivated area is followed by declining productivity rates.

There is also a distinct variation in the rates of growth in cultivated areas for each crop in the different regions and periods. The growth rate of food crops is much higher on the agricultural frontier (North, Central-West and part of the South) and in traditional regions of export crops in decadence (North-East). These figures show that a major part of food production has continued to be neglected, still being provided by small farmers, tenants, share-croppers, and *posseiros*[2] on the agricultural frontier, on the boundaries between large properties or on the plantations where export production decreased because of the world market depression, but could rapidly be expanded given more favourable conditions. In spite of this, agricultural production was no obstacle to the industrialization process, either in producing enough food surplus or enough raw materials for industry. But this does not imply that there was no problem of supply in the urban areas. Such a food production structure has continued to reinforce the power of the *atravessadores* to keep food prices high without modifying the price relation between agriculture and industry (as we see in Table 4.4), except for periods of crisis in supply when the government usually imposed price controls on some essential products. But the agricultural sector as a whole in this period did not represent an obstacle to industrial growth, as Rangel (1963) and others have supposed.

The high price of food had no impact upon industrial production costs, but lowered the quality of life of the labouring classes, who had to spend a great deal of their income on food. The highly concentrated landownership structure together with the strong demographic growth created a surplus of population which migrated, for the major part, to the cities, thus influencing urban wages. On the other hand, a large part of the equipment adopted in industry was labour saving, imported from countries where labour was scarce and the unions strong. As a result wages remained low during the whole period of fast industrial growth. The major portion of productivity gains were appropriated by the capitalists, resulting in a very unequal pattern of income distribution, as in the agricultural sector. In this sense we can say that agriculture projected 'its image on the urban-industrial sector' (Castro, 1975, vol. 1, p. 75), helping to determine a structure of production polarized between a traditional sector producing wage goods for the low-income majority,

Table 4.4   Price indexes: average yearly growth rates, Guanabara state, 1948–72.

Period	General	Food	Clothes	Habitation	Household goods	Public medical assistance	Personal services	Public services
1948–50	6.7	6.8	4.3	10.7	0.8	11.3	9.4	10.5
1950–54	16.5	18.1	12.0	19.1	10.5	6.5	10.7	11.3
1954–58	18.3	19.4	15.4	16.8	17.3	20.5	17.8	27.7
1958–62	38.3	43.0	40.7	23.1	40.5	38.8	46.7	35.0
1962–66	67.4	61.9	65.6	69.1	70.7	66.2	74.4	89.8
1966–70	24.4	21.0	22.9	33.6	22.1	26.5	28.2	26.0
1971	20.2	22.5	16.8	16.8	15.8	21.5	20.8	24.0
1972	16.8	16.8	12.7	9.8	8.8	14.9	20.8	23.7

Source: Paiva, R. M., S. Schatten and C. F. T. Freites 1973.

and a dynamic sector producing durable goods for a relatively small high-income market.

In this period there was no internal market for the agricultural inputs and machinery produced by the industrial sector, because of the low technology generally adopted in agriculture. But this was no barrier to the industrialization process either, and Brazilian agro-industry was established by the late 1950s. Agriculture also helped to finance the industrialization process, as financial resources were transferred to the industrial sector by keeping constant the rate of exchange in an inflationary period, that meant an overvaluation of the national money *vis-à-vis* US dollars, this favoured imports (which, in turn, were subject to a selective control that favoured capital goods, industrial inputs and raw materials for the industrial sector) by transferring resources from exports which had been mainly composed of agricultural products (Furtado 1963). In short, the diminishing rate of industrial growth at the beginning of the 1960s was not caused by the performance and characteristics of the agricultural sector. The huge amount of land available, a rapidly expanding agricultural frontier and a relatively low population density permitted the agrarian structure to remain unchanged in spite of the shift of political power to the urban-industrial elites.

It should be noted that the reorganization of the economy after the military *coup d'état* in 1964 was made by greatly reinforcing and enlarging the industrial growth pattern, then going through its first crisis. The basic wage, which had been slowly increasing, was lowered, forcing women and children on to the labour market to complement a family income that had become insufficient even to provide a very low standard of living. Meanwhile, following this reduction of the basic real wage, the government enlarged the buying power of the upper class and consumer credit facilities in order to increase the market for the dynamic sector of the economy (durable goods), dominated by multinational enterprises.

Crisis and change

The agricultural sector underwent changes to tenancy arrangements and the level of technology in the period following the crisis of the early 1960s, but the landownership concentration remained the same. To understand these changes we must consider, first of all, one of the main characteristics of the Brazilian agrarian structure: the use of land as a reserve of value. Because of speculation, buying land became a sure way of protecting savings against inflation. The capital invested in land is profitable independently of new investments to make it productive. In addition, the policy of subsidized credit and tax exemptions was based only on land area, favouring large properties. So, besides being a reserve of value, land also provided access to other forms of wealth. As a result there is a high rate of unused land, mainly in the most developed regions. Table 4.5 clearly illustrates this situation in the most developed state, São Paulo. As we see in the first column, there is a clear correlation between the size of the property and its agricultural production value. In other words, the percentage of properties with a production value up to 3 cruzeiros falls as size rises. But for properties of 500–1000 hectares this tendency is reversed; the percentage of properties with more than 1000 hectares and a production value of up to 3 cruzeiros rises significantly, with 41.5% of farms of more than 10 000 hectares in this category. These figures show the enormous waste of land on the large farms. In addition, some properties are completely unexploited (last column), literally abandoned, waiting for land prices to rise: around 21% of the properties with more than 10 000 hectares, and 21% of those with less than 2 hectares – the latter probably consisting of plots around the urban centres (Graziano da Silva 1980).

At the beginning of the 1960s important changes began to take place in the prevailing relationships between landowners and landless peasants. The traditional patron–client tenancy arrangements – a personal relationship where the landowner guaranteed, for instance, land for peasants to cultivate their food in exchange for cheap labour[3] – allowed the landlords a large measure of political control over the peasants. This kind of tenancy arrangement was undermined by two events: on the one hand, the growing political weight of social reform movements urging agrarian reform (as we mentioned above, the latifundio was seen as an obstacle to the process

Table 4.5 Distribution of rural establishments by area and value of production per farm, São Paulo state, 1972 (percentages).

Area (ha)	Value of production[a] (cr $1000)					Total	Unexploited establishments[b]
	< 3	3–12	12–50	50–100	> 100		
< 2	89.3	6.5	2.8	0.8	0.6	100.0	21.1
2–5	74.3	17.5	5.9	1.4	0.9	100.0	10.5
5–10	57.3	30.8	9.5	1.3	1.0	100.0	6.5
10–25	37.7	39.9	20.0	1.8	0.8	100.0	4.3
25–50	26.2	36.7	31.4	4.2	1.3	100.0	3.4
50–100	19.3	29.2	40.0	8.2	3.4	100.0	3.1
100–200	14.3	18.9	42.0	14.9	10.0	100.0	3.0
200–500	12.1	9.6	31.1	20.1	27.0	100.0	3.5
500–1000	11.2	4.6	16.1	15.7	52.4	100.0	3.9
1000–2000	12.9	2.5	7.8	10.3	66.2	100.0	4.7
2000–5000	15.5	2.5	5.2	4.9	71.9	100.0	5.7
5000–10 000	21.9	3.1	3.9	3.1	68.0	100.0	6.8
> 10 000	41.5	2.4	–	–	56.1	100.0	20.9

Source: Graziano da Silva 1980.
[a] Excluding establishments non-classified by total area or production value.
[b] These establishments are either unexploited, or where there is no labour force, production value is not recorded and non-agricultural activities do not take place.

of industrialization and economic development, and so its legitimacy was questioned); on the other hand, the rapid growth all over the country of unions of small farmers and landless peasants. The government tried to control and direct these movements by creating a new law, *Estatuto do Trabalhador Rural* (Law 4214, 2 March 1963), which attempted to organize the rural unions in the same way as urban unions (attached to the government), and took away the right to strike (Sorj 1980). As a compensation for the loss of the union's freedom, social legislation was extended to the peasants, which meant a rise in labour costs for the landowners. In spite of the reactionary military *coup d'état* in 1964 with the resulting repression of the workers and suppression of free unions, it was the two events outlined above which caused the definitive break with traditional tenancy arrangements.[4]

The consensus at all levels of society about the unjust and miserable situation of the peasants, and the need to extend to them the social legislation given to urban workers, combined with the arousal of a political consciousness among the peasants themselves, became a serious obstacle to traditional land speculation. The existence on the farms of workers more or less protected by the law compromised land as a liquid asset. In the state of São Paulo during the 1960s it was stated explicitly in a large number of land transactions that the seller was responsible for the labour indemnities of those living on the property (Graziano da Silva 1980). The landowners reacted by chasing off the workers and substituting their food crops with pasture, the traditional way of controlling land without much labour input. The peasants became temporary wage workers (*volantes*) living on the outskirts of cities. They were recruited by intermediaries (*gatos*) who transported them by trucks to the farms. These workers had no legal rights, medical assistance, retirement benefits and so on; they earned such miserable wages that they were forced to send their children to work as young as nine years old. This proletarianization process of rural workers has accelerated greatly since 1967 as a result of agricultural modernization. Partial mechanization in the most important crops means a greater seasonality in labour demand; partial mechanization concentrates labour demand at certain phases of the production process: for crops such as sugar cane, coffee, oranges and cotton there are still unsolved technical difficulties in the mechanization of harvesting. As a result, there are longer periods without work during the different phases of the production process which would still have to be paid for if the workers lived on the farms.

Besides these modifications in the tenancy arrangements there was also, in this period, a general awareness of the need to raise the technical level of agriculture; land productivity was very low and tended to fall. Obviously, the existence of an industrial complex producing agricultural inputs and machinery contributed to this awareness, had an effect on the kind of techniques to be adopted. It should be noted that many of those industries were multinational enterprises producing in Brazil the same products as in their own countries, where the ecological factors and human resources might be completely different.

Trends in agricultural production in the period (1963–7) were similar to those of the previous period, with food production and export crops growing sufficiently. The rapid expansion of soybean production, which was to have a great impact on the structure of landownership, production and employment in the southern part of the country, should also be pointed out.

The recent expansion of Brazilian agriculture

Agriculture modernization, rural exodus and food scarcity

In this new period of accelerated industrial growth, agriculture played an important role as a source of currency and as a market for

Table 4.6 Yearly growth rates of principal crops in Brazil, 1947–67 and 1967–79 (percentages).

Crops	1947–67	1967–79
Internal market		
rice	5.96	1.94
potatoes	4.39	2.96
beans	4.05	−1.26
cassava	4.80	−1.51
corn	4.74	2.50
Average	4.79	0.93
Export		
cotton	3.79	−2.18
cacao	1.79	3.99
coffee	4.12	−1.56
sugar cane	5.82	5.57
oranges	4.60	14.88
soybean[a]	14.33	27.38
Average	5.73	8.01

Source: Graziano Neto, FO 1982.
[a] For soybean the first period considered was 1951–67.

industrial products. It was needed as a currency source, because not only did the new industrial expansion require a higher level of imports, but also private investments by multinational and national (mainly government) enterprises caused a heavy external debt, and therefore exports had to be increased in order to pay for imports and to cover the currency deficit of the balance of payments. It was needed as a market for industrial products because of the existing capacity of production of the agro-industrial complex installed by the end of the 1950s. Food production for the internal market continued to be ignored; the proposals for agrarian reform were dead and buried. As we can see in Table 4.6, the rate of growth of the main food crops between 1967 and 1979 are substantially lower than the rates in the preceding period (1947–67). As for export crops, the rates of growth vary widely depending on the crop. This variation can be explained by exceptionally favourable conditions in the world market for some crops (soybean, oranges) and poor conditions for others, together with internal production problems (coffee).

Table 4.7 shows clearly that there was an expansion of export crops at the expense of food crops between 1967 and 1979 in the southern and south-eastern regions (where the major part of Brazil-

Table 4.7 Division of cultivated area into food and export crops, south and south-east Brazil, 1967 and 1979.

Crops	1967		1979	
	Area (1000 ha)	Percent	Area (1000 ha)	Percent
food crops	11.900	63.0	16.100	55.5
export crops	6.900	37.0	13.000	44.5
total	18.800	100.0	29.100	100.0

Source: Menezes *et al.*, 1981 in Paschoal 1983a.

Table 4.8 Number of tractors in Brazilian agriculture, 1950–80

Year	Tractors (units)	Cultivated area per tractor (hectares)	Establishments per tractor
1950	8 372	2,281	247
1960	61 338	468	54
1970	165 870	205	30
1975	323 113	124	15
1980	527 906	87	10

Source: Graziano Neto, F. 1982.

ian agriculture takes place). The percentage of land under food crops in the total agricultural area fell from 63% to 55%. When the crops are taken individually, soybeans are definitely mainly responsible for this modification. The soybean cultivated area in the states of Paraná and Rio Grande do Sul – the main producers – leapt from 577 000 hectares in 1967 to 6 450 000 hectares in 1979 (Paschoal 1983a). Previously, an expansion in export crops at the expense of food crops was balanced by an increase in the production of the latter in other regions, or in the same region on less fertile lands. This time, however, the increased growth of crops, the pushing forward of the agricultural frontiers to barely accessible regions and the expulsion of the workers living on the farm all compromised food production for the internal market. In addition, the expansion of soybeans was mainly in the regions where a polycultural peasant agriculture based on small farms had been developed. After the second oil crisis (in 1978) and following government incentives, the expansion of sugar cane for the production of alcohol as a substitute fuel for cars, has competed with food crops in the state of São Paulo.

Together with the rapid expansion of export crops, the moderni-

Table 4.9 Consumption and growth rates of fertilizer in the principal commercial crops,[a] Brazil, 1960–77.

Year	Consumption of mineral nutrients (tons)	Growth index (1960 = 100) (1)	Growth index of cultivated area (1960 = 100) (2)	Growth index of fertilizer use per hectare (1/2)
1960	298 734	100	100	100
1961	247 177	82	103	80
1962	236 875	79	108	73
1963	314 044	105	112	93
1964	255 245	85	114	75
1965	209 399	97	122	79
1966	281 119	94	117	80
1967	447 925	150	118	127
1968	601 708	201	122	165
1969	603 385	211	127	165
1970	999 040	344	135	248
1971	1 165 036	390	129	302
1972	1 746 525	562	130	448
1973	1 679 147	548	133	421
1974	1 824 636	611	147	415
1975	1 977 672	662	151	424
1976	2 528 143	846	163	519
1977	3 149 068	1054	172	613

Source: Agricultural Census and Statistical Yearbooks, FIBGE; Union of Agricultural Fertilizers Industry of São Paulo.

[a] Rice, cotton, coffee, sugar cane, corn, soybean and wheat.

zation of agriculture accelerated greatly. The number of tractors (Table 4.8) grew enormously, reaching half a million in 1980, and the ratio of cultivated area to tractors fell to one tractor for every 87 hectares. The consumption of fertilizers, steady at about 290 000 tons (of pure mineral nutrients) between 1960 and 1966, jumped to 3.15 million tons in 1977. If we discount the expansion in the cultivated area of the seven main crops (cotton, rice, coffee, sugar cane, corn, soybean and wheat, in the states of São Paulo, Paraná and Rio Grande do Sul) which consume 65% of the fertilizers in Brazil, we will have an approximate measure of the growth of fertilizer use per hectare (Table 4.9). The amount of fertilizer per hectare grew five times in this period, reaching, in 1977, about 300 kg. Similarly, pesticide consumption grew rapidly, multiplying five times since 1964 (Table 4.10); in those regions it reaches equivalent amounts per hectare to those used in England and Japan (Paschoal 1983b).

Table 4.10 Pesticide consumption in Brazilian agriculture, 1964–78.

Year	Consumption (tons)			
	Total	Insecticides	Fungicides	Herbicides
1964	16 193	12 560	3 268	365
1965	22 393	17 932	4 220	241
1966	30 241	22 444	7 342	455
1967	25 455	16 475	6 059	921
1968	35 943	28 465	5 647	1 831
1969	40 650	33 514	5 685	1 451
1970	39 469	28 306	7 747	3 416
1971	43 744	27 223	11 514	5 037
1972	63 485	33 899	24 698	4 888
1973	84 311	37 898	36 945	9 468
1974	100 219	45 247	40 533	14 439
1975	77 083	41 803	13 892	21 388
1976	68 214	28 500	16 357	23 357
1977	78 357	33 846	24 585	19 926
1978	81 447	39 985	18 461	23 001

Source: Union of Agricultural Pesticides Industry of São Paulo.

This process of modernization was stimulated and supported by an agricultural policy of expanding subsidized credit. Between 1970 and 1980 the number of credit allowed contracts in the National Rural Credit System grew by 132%. In money terms, credit allowed grew from 9.2 billion cruzeiros in 1970 to 859.2 billion in 1980, a real growth of about 87%. The implicit subsidies represented about 35% of the total amount of credit granted (Munhoz 1982).

This new phase of Brazilian agricultural expansion has caused many problems. The first is in food production. As we have seen, the rate of growth of food crops fell noticeably in the period 1967–79. As a result the government was constantly forced to import in order to guarantee provisions for the domestic market. From about 1978, the government has been trying to stimulate food production for the domestic market by increasing credit for corn, beans and rice (the Brazilian staple food). In 1970 these crops had a ratio of financed cultivated area to ordinary cultivated area of about 25.4%, 17.4% and 29.8% respectively; in 1980 these ratios grew to 57.8%, 78.5% and 84.4% (Munhoz 1982).

The second problem is unemployment. Increasing mechanization together with the expansion of crops like soybean and wheat that are mechanized in all phases of the production process have greatly accelerated the rural exodus to the cities. As well as landless peasants being thrown off large farms, there were also an important number of small farmers who lost their lands. As we saw above, soybean expansion (and that of wheat, cultivated in rotation with soybean) took place mainly in the regions of small farmers who were obliged to become 'moderns' (contractual clauses linked the credit concessions to the purchase of agricultural machinery and modern inputs); they began to have increasing financial problems as they experienced greater difficulties than the big farmers in obtaining subsidized credit during the bad years (Pearse's talents effect: the big farmers are better educated and better informed and are therefore more able to take advantage of the agricultural policy), and also to problems caused by scale of production of the equipment and the rising costs of inputs (fertilizers, pesticides, and so on) after the oil crisis.[5] These growing financial difficulties on the one hand, and the rising land prices on the other, forced the small farmer to sell his property. If after paying his debts there was enough money left, he might move to the agricultural frontier where, because of the big difference in land prices between the two regions, he might be able to start farming again. If not, this expropriated small farmer simply joined the underemployed contingent that fills the slums on the outskirts of the big cities.

Thus in the 1970s landownership became even more concentrated. It is an important to note that in this process the degree of land use fell, showing very clearly the heavy land speculation that follows, historically, agricultural expansion in Brazil. This is shown by the National Institute for Colonization and Agrarian Reform (INCRA) in the figures for the state of Paraná, the largest national producer of grain: in 1972 the participation of agricultural establishments classified as *rural enterprises* (that are exploited *rationally* and *economically*, according to INCRA) in the total agricultural area was about 17.9%, falling to 13.3% in 1978; at the same time the participation of agricultural establishments classified as *latifundio by exploitation* (agricultural units that, irrespective of size, are unused or used below their potential) rose from 58.1% in 1972 to 63.8% in 1978 (IPARDES 1981).

The figures in Table 4.11 illustrate the modifications taking place in the agricultural production structure with regard to the type of producer. The number of establishments, as well as the area they covered, where the owners are the direct producers grew between 1970 and 1975 for both temporary and permanent crops; in the same period both numbers and percentage of tenants and sharecroppers fell whereas occupants rose (the majority being *posseiros*). These figures reflect the expulsion process, started in the early 1960s, of the landless workers living on the big farms (tenants and sharecroppers)

Table 4.11 Agricultural establishments according to type of producer, Brazil, 1970–5.

Producer	Permanent crops				Temporary crops			
	Establishments		Area (ha)		Establishments		Area (ha)	
	1970	1975	1970	1975	1970	1975	1970	1975
owners	1 132 332	1 153 977	6 608 107	7 298 505	2 604 409	2 737 207	19 495 569	24 863 256
tenants	76 260	54 722	249 450	173 685	586 887	538 295	2 834 881	2 930 361
sharecroppers	125 056	89 374	706 948	450 774	321 102	260 033	1 564 470	1 393 869
occupants	166 812	186 733	419 562	462 431	730 656	847 557	2 104 807	2 428 477
total	1 500 460	1 484 806	7 984 068	8 385 395	4 243 054	4 383 092	25 999 728	31 615 963

Source: Statistical Yearbook 1981, FIBGE – Brazilian Institute of Geography and Statistics Federation.

who became temporary wage workers or *posseiros*. In the 1970s this expulsion process was accelerated by the expansion of export crops together with rapid, labour-saving mechanization. This fact added to the demographic growth and the establishment of large cattle farms on the agricultural frontiers (stimulated by the government through tax exemptions)[6] greatly increased the incidence of (usually armed) struggles over possession of land which exploded all over the country, not just on the agricultural frontiers. The government intervened (with the army in certain regions) generally on the side of landowners. This growth in conflicts during the 1970s emphasized the dramatic situation of the Brazilian peasant: poor, no guarantee of land tenure, frequently expelled by force in spite of there being large tracts of abandoned land, and often to make way for extensive cattle-breeding operations which provide a more profitable way of controlling large areas of land. According to INCRA there were about 80 million hectares of land classified by the owners[7] as non-arable land in 1975, and 32 million hectares classified as unexploited, abandoned until prices rise. Subtracting the unexploited and non-arable areas and the cultivated area (40 million ha) from the total agricultural area (323 million ha), we are left with 170 million hectares devoted to livestock, mainly cattle, for a herd estimated at 100 million head, giving a ratio of 0.58 animals per hectare!

Environment and agricultural modernization

The 'ideal' modernization pattern stimulated by the government mainly through credit policies is, as we have seen, a reproduction of the American and European model. This is considered to be the only way to raise the technical level of agricultural practices in the country. The agricultural equipment and input industry has certainly played an important role in determining the characteristics of the process of modernization. In 1973 EMBRAPA (Brazilian Enterprise of Agricultural Research) was created to co-ordinate the country's agricultural research, which was concentrated in a small number of areas (mainly export crops), as in the big international institutes of agricultural research (IRRI, CIMMYT, and so on) responsible for the so-called 'Green Revolution'. The transfer from outside of a technological package is seen as a way to 'capitalize on the investments already made in other countries'. The private sector must work as an efficient 'guide and controller of the major part of research projects, and extension services will need to be developed in close relationship with the agricultural input industry' (Pastore & Alves 1975).

Questions as to the probable inadequacy for a country of tropical climate of agricultural techniques adopted in countries of temperate climate are not even considered. The consequences of this technological transfer from an environmental point of view could not be worse. The system of soil preparation used in temperate countries has a devastating effect, in erosion terms, when applied to tropical countries. In temperate regions, to prepare the soil for planting it is necessary to turn it, exposing it to sunlight, so that by raising its temperature (in such regions to a maximum of 18 °C) its life (bacteria, worms, and so on) is reactivated. In the tropics the problem is literally the opposite, for soil temperature can reach 75 °C, which is hot enough to burn all the life out of unprotected soil (Primavesi 1980). Enormous erosion is caused by exposing unprotected soil, pulverized by heavy ploughing and harrowing; in addition, rows of monocultural crops to permit machine operations offer very little protection in a region of heavy and unevenly distributed rainfall.

Research by the Agronomic Institute of Campinas (IAC) has concluded that in Brazil average soil loss is greater than 25 tons per hectare per year. Measures from the Ribeirão do Rato area, northwest of the state of Paraná, showed soil loss between 22 and 187 tons per hectare per year. In the regions of highly mechanized temporary monoculture such as soybeans, the soil loss is estimated at 100 tons per hectare per year, while the limits of tolerance calculated by IAC are between 4.5 and 15 tons (Bertoni *et al.* 1972) – more or less the same figures estimated for the USA by the Soil Conservation Service and Soil Survey Staff. The disastrous results of this situation become obvious when we realize that the state of Paraná loses about 1 cm of fertile soil per year (Scroccaro 1980), a loss it would take around 400 years for nature to replace. The situation is so dramatic that the co-ordinator of PROICS (Integrated Programme of Soil Conservation) in Paraná proposes as an acceptable goal the reduction of the loss to 25 tons per hectare per year (Mazuchowski 1980).

Chemical fertilizers are used to try to compensate for the deficit of nutrients caused by erosion. Consequently a large amount of chemical fertilizer is lost with the soils. Figures for the soybean regions in the state of Rio Grande do Sul estimated by EMERAPA show the huge loss of mineral nutrients caused by erosion: 147 000 tons of phosphorus out of 380 000 tons added and 210 000 tons of potassium of 202 000 tons added in one year. In the latter case there is a net loss of potassium in the soil (Tomasini, Wunsche & Portella 1981). This violent erosion process will obviously have a negative impact in yield per hectare. Tables 4.12 and 4.13 show the low productivity growth of the seven main commercial crops in São

Table 4.12 Yield per hectare of the principal commercial crops in the states of São Paulo, Paraná, and Rio Grande do Sul, 1966–77.

Year	Cotton		Rice		Sugar cane		Coffee		Corn		Soybean		Wheat	
	Kg/ha	Index	Kg/ha	Index	Kg/ha	Index	Kg/ha	Index	Kg/ha	Index	Kg/ha	Index	Kg/ha	Index
1966[a]	1004	100	1818	100	54755	100	1077	100	1581	100	1195	100	785	100
1967	1147	114	1760	97	54501	99	955	88	1570	99	1092	91	829	105
1968	1191	118	1737	95	52316	96	1072	99	1559	98	1080	90	871	111
1969	1227	122	1775	98	52791	96	797	74	1389	88	1077	90	951	121
1970	1208	120	1842	101	52881	98	848	78	1376	87	1184	99	976	124
1971[b]	1237	123	1905	105	53922	98	806	75	1385	87	1227	102	968	123
1972[b]	1296	129	1874	103	53408	97	1070	99	1608	101	1307	109	1014	129
1973	1344	134	1927	106	53468	97	1263	117	1676	106	1423	119	1088	138
1974	1367	136	1988	109	53984	98	1349	125	1761	111	1573	131	974	124
1975	1431	142	2086	115	56359	103	1120	104	1840	116	1475	123	906	115
1976	1502	149	2179	120	59807	109	907	84	1931	122	1557	130	731	93
1977	1412	140	2143	118	63115	115	864	80	1790	113	1397	117	844	107

Source: Agricultural Census and Statistical Yearbooks, FIBGE.
[a] Figures are three-year averages centring on the years shown.
[b] Estimated figures.

Table 4.13   Yield per hectare indexes of principal crops, São Paulo state, 1967–79 (1966–8 = 100)

Years	Cotton	Rice	Coffee	Sugar cane	Corn	Wheat	Soybean	Orange
1966–68	100	100	100	100	100	100	100	100
1969–71	82	92	111	94	95	135	85	76
1972–74	82	135	146	94	116	171	111	80
1975–77	102	128	91	94	120	82	123	82
1978–80	100	115	108	100	120	119	116	100

Source: Graziano Neto 1982.

Paulo, Paraná and Rio Grande do Sul (the most important and modernized agricultural areas) during 12 years of accelerated modernization (1966–77), a period in which the quantities of fertilizers applied per hectare increased five times, as we saw in Table 4.9 above.

In a process analogous to the one by which the transfer of soil handling techniques from cold countries has generated heavy erosion in tropical climates, the spread of chemical control of pests has aggravated them in Brazilian agriculture, besides having undesirable effects on human health. The ecological conditions of the tropical regions are obviously completely different from those of the temperate climates. In the tropics the temperature varies little throughout the year and the photoperiod is very stable. Therefore, the main factors of stability and equilibrium are biological, not physical (as is cold and snow in temperate regions), and result from a huge diversity of life forms in complex interrelationships (Paschoal 1983b). In this kind of ecosystem the chemical control of pests is inefficient, since it destroys the complex system of equilibrium. The natural predators are eliminated while the pests develop resistance to the pesticides, through a well-known process of genetic mutation. Certain species previously considered inoffensive begin to reproduce uncontrollably, also becoming harmful.

In short, the general decay of the agricultural environment in Brazil, resulting from more than four centuries of a predatory concept of farming, is greatly accelerated by the modernization process. In the face of exhaustion of the soil the farmers substituted natural soil nutrients with chemical fertilizers. In this sense, we can say that 'modern' agriculture in Brazil, more than anywhere else in the world, 'transforms renewable resources such as a "living" soil to non-renewable resources; each harvest becomes an "assault" on nature (fuels and soils) and no longer a product of a natural cycle of organic material production' (Romeiro & Abrantes 1981).

The alternative developments

The technological choice criteria

Alternative developments in Brazilian agriculture must take into account three fundamental criteria: (a) compatibility with labour availability; (b) energy conservation; (c) ecological sustainability. The first suggests relatively labour-intensive technologies for the majority of developing countries. However, the increase of labour input per hectare must be correlated with maintaining a labour productivity which allows the peasant a reasonable standard of living. For the second criterion, the reduction in consumption of non-renewable energy in farming processes must be reached without reducing the yield per hectare, which means an energy-efficient agricultural ecosystem. Finally, the need to preserve natural resources is not incompatible with high levels of land productivity; there is clear evidence on this.

Many people think that the application of these criteria would mean a return to the past: the return to labour-intensive agricultural practices which are degrading and undignified for the human condition; to energy-saving and ecologically sustainable practices giving low yield per hectare. They believe the current agricultural modernization process to be the only answer to the need for raw materials and food in the contemporary world, although they accept its attack on the environment and its dependence upon non-renewable energy sources. Here there are two basic standpoints, varying only in being optimistic or pessimistic about the chances of mankind's survival faced with the exhaustion of natural resources and the degradation of the environment. On one side there are those who are truly faithful to the ability of technical progress, such as it is now, to solve all the problems created by technical progress itself. Considered as a *deus ex machina* it will always be possible for technical progress to solve or diminish problems caused by the degradation of the environment and to find new energy sources without modifying the course of modernization (Schultz 1974). On the other side there are the fatalists who believe that, unless the world population returns to the levels of 1000 years ago and practices a balanced subsistence agriculture, mankind is marching inexorably towards its own destruction. It will be necessary to use all fossil fuels for agriculture and look for other energy alternatives for urban-industrial activities in order to prolong the life expectation of mankind (Georgescu-Roegen 1975).

In our opinion these conceptions are false. We do not think that ecologically equilibrated agricultural techniques belong to the past. There are certainly some procedures and techniques developed by traditional peasant agriculture (in Europe and Asia) which can still be of use today. However, what we have in mind is not the return to traditional agricultural practices themselves, but to their rationality. It is a matter of bringing this rationality to another level of scientific and technological knowledge (Sachs 1980). Technological and scientific development today make it possible to create highly productive agricultural systems which preserve the equilibrated and self-sustained structures of traditional peasant agriculture. The development of agricultural systems in which scientific and technological knowledge can act as a 'factor' of production in place of land, labour and capital is what we have in mind.

In the current modernization process the diffusion of technological knowledge is mainly embodied in inputs, machines and agricultural equipment whose use has precise technical directions which do not require the farmer to have a close familiarity with his own environment. He only needs to follow the instructions and prescriptions of the technical advisers correctly. Consequently much potential in the handling of available natural resources is lost. The development of a technology more suitable to each ecosystem is substituted by a patterned technology embodied in machines and agricultural equipment. Some examples of how scientific knowledge can work as a production factor in place of capital, land and labour are given below:

(1) Scientific knowledge can be used in place of capital and land in (a) the handling of life in the soil (bacteria, worms, and so on), in order to obtain adequate soil texture, as a substitute for mechanical tilth; an increase in the capacity of the soil to retain and store water, in place of irrigation; and the recycling and production of mineral nutrients as a substitute for chemical fertilizers; (b) the genetic selection of plant varieties tolerant to adverse soil factors like mineral and water stress (see Swaminathan 1981).
(2) Scientific knowledge can be used in place of labour and capital in the biological control of pests and weeds instead of chemical pesticides and labour-intensive weeding.

Thus the ideal agricultural production function must be capital-saving, land-saving, relatively labour-intensive (depending on the specific conditions of each region) and intensive in scientific knowledge. Biotechnology and computing open huge perspectives for a better exploitation of the potential offered by the environment in terms of pest control and vegetal nutrition. This means augmenting the use of solar energy in agricultural production. It is a matter of creating agricultural systems where the chemical fertilizers are used as correctives and supplements rather than as the principal sources of vegetal nutrition; and where the chemical control of pests, if necessary, is prompt and not permanent.

The agricultural sector, unlike the industrial sector, presents many possibilities for combining land, capital and labour. However, these multiple combinations are not neutral from the point of view of preserving the environment and its natural resources. Generally, in the analysis of the productive process in agriculture a production function, similar to that used to study the production process in the industrial sector, is used, the only difference being the addition of another production factor, land. The proportions of these factors vary in the same way as those of the industrial sector, according to the local availability of production factors. So historically a progressive augmentation in capital has been observed as a result of the growing scarcity of labour and land (Hayami & Ruttan 1971). However, the substitution of labour and land by capital can cause, as has been happening, a higher deterioration in land. In American agriculture, where labour productivity has reached incomparable levels, land has been progressively degraded (see Rodale Press 1981). It is impossible to expect that it will have an equivalent time horizon to the 2000 years of European agriculture or the 5000 years of Chinese. The soils are compacted by the use of heavy machines; erosion is accelerated through a system of planting crops in rows to simplify mechanization, which leaves the soil unprotected from erosion (mainly by wind on the huge plains); the soil and water courses are polluted by chemical fertilizers and pesticides.

In Brazil a greater preoccupation with the environmental problems caused by 'modern' agricultural practices is evident, along with research into alternatives which are better equilibrated ecologically and more energy efficient. Since the 1972 oil crisis the cost of agricultural inputs have risen in spite of the subsidies. In addition, the amount of inputs (fertilizers and pesticides) per hectare has also risen to keep up productivity levels, as we saw above. In consequence of soil decay and the rising cost of recuperating it, techniques of zero or minimum tillage, for instance, seem to be expanding. There are many similar techniques which need not modify the agrarian structure. It seems that the limit for the adoption of techniques which are more equilibrated in ecological terms is set by the greater amount of labour they require; mixed crops, for example, make it difficult or even impossible to mechanize the harvest, but landowners do not wish to take on more workers.

The labour absorption problem

What we have seen so far of the process of agricultural modernization in Brazil is still not enough to explain the acceleration of labour-saving mechanization in the midst of abundant cheap labour. From the point of view of land productivity, the chemical and biological innovations that characterize modern agriculture (such as high-yielding plant varieties) are neutral with regard to labour use; it is the work complementary to their use which may or may not be labour-saving. The process of agricultural modernization in Japan illustrates this well. The vigorous modernization and expansion that began at the end of the 19th century, in the Meiji era, was carried out with increasing amounts of labour per hectare in the first phase (see Ishikawa 1981). Hayami and Ruttan (1971) saw in this process a good example of their induced innovation theory. For them, technical innovation in agriculture is encouraged by the relative production factor endowments. Thus in Japan, where there was land scarcity and labour abundance, agricultural techniques tended to be labour intensive and land saving. The USA represents the opposite case. However, agricultural modernization in the majority of developing countries was followed by an acceleration of labour-saving mechanization, in regions where huge peasant masses had no chance of being absorbed into the expanding urban–industrial sector. In this sense, the Japanese case seems to be the exception that confirms the rule.[8] But what rule? This is the question we are going to look at next.

First, it is necessary to take into account a common fact that links countries as different as Brazil, India and Indonesia: the monopoly of the land by a class of owners who are indifferent to the fate of peasants dependent on it for survival. Historically, relations between the agrarian elites and the peasantry has assumed different forms according to each culture, but in all the forms there was a mixture of open violence and personal relationships (involving the compromise of guaranteeing for the peasants a minimum level of security for their survival).[9] This kind of relationship allowed the landowner a good organization and control of the working process with a large number of workers. When this kind of tenancy arrangement is substituted, for several reasons, by capitalistic

labour–wage relationships which end personal relations between
landowners and landless peasants based on traditional community
principles, the organization and control of the working process
becomes more difficult. The seasonal character of agricultural
activity, the discontinuity of labour demands in different phases of
the production process, and the attention and care demanded from
the worker in several operations causes managerial and supervisory
problems and difficulties in hiring labour during peak seasons. In
certain regions, even if labour-saving equipment is more expensive
than labour, its higher costs are compensated for by the increased
leisure and better working conditions obtained (see Pearse 1980).
However, in all developing countries where agriculture is being
modernized, agricultural policy subsidizes labour-saving machinery
and equipment. So the relative prices of capital and labour are
modified by agricultural policy to benefit the former rather than the
landless peasant.

In Brazil, labour-saving mechanization is justified by the policy
makers as a 'natural' consequence of the rising cost of wages.
Hayami and Ruttan's model of induced technological innovation is
used to rationalize this process in terms of economic theory (see
Alves 1981).[10] Thus we can suppose that the 'high wages' of the
miserable mass of peasants constituting the labour force in Brazilian
agriculture are responsible for the introduction of capital-intensive
agricultural techniques. Sanders and Ruttan (1978) admit 'dis-
tortions' in Brazilian agricultural policy (through subsidies) to
explain this labour-saving mechanization in a country of labour
abundance. The large farmers from the south of the country are to
blame, as they were able to influence the government to subsidize
mechanical technology as a substitute for labour in order to shape
technical change to release the constraints of their relative factor
endowments (and so the model of induced innovation is preserved).
This assumes that rural wages in the south would be rising faster
than the prices received by the farmers, but Table 4.14 shows that
this is not the case. The first column shows the price indexes of
agricultural products in the state of São Paulo between 1968 and
1977. If we compare them to wages we can verify that for resident
workers wages grew at a rate consistently lower than prices, and
while wages of temporary workers varied, there was no tendency
for them to rise above agricultural prices. Prices of fertilizers and
tractors clearly reflect the strong subsidies they received. Real prices
of tractors fell steadily throughout the period. The fast growth of
land prices (last column) should also be noted, resulting from the
speculation which, as we saw, has always followed agricultural
expansion in Brazil.

Table 4.14 Real price indexes[a] of selected agricultural components, São Paulo state, 1967–77 (1967 = 100).

Year	Agricultural products[b]	Chemical fertilizers	Tractors (44 hp)	Wages		Agricultural land
				Resident workers	Temporary workers	
1968	101	98	97	106	118	111
1969	120	95	94	103	109	83
1970	117	86	78	114	124	103
1971	118	90	75	119	128	124
1972	134	95	72	132	146	135
1973	174	102	67	156	161	194
1974	157	208	58	169	193	397
1975	171	114[c]	60	169	193	388
1976	227	85[c]	67	161	194	427
1977	241	129	71	172	210	427

Source: Graziano da Silva 1980.
[a] Deflator: index 2 FGV (Getulio Vargos Foundation).
[b] Price indexes received by the farmers.
[c] Years when a direct subsidy of 40% above the selling value was instituted.

In short, to understand the introduction of labour-saving mechanization in countries with abundant supplies of labour, it is not sufficient to take into account the power of the agro-industrial 'lobby' and its influence on agricultural policy and farmers through its efficient marketing technical assistance network. Mechanization is of great interest to the large farmers who control most of the agricultural area of the country, not because of rising wages but because of the particularities of the agricultural production process that present, on a strictly capitalistic basis, insoluble problems for the organization and control of the working process. In other words, the cost of labour as given by nominal wages is not important to explain the 'take-off' of the labour-saving mechanization process in Brazilian agriculture. To do so, we need to consider the cost of labour as given by deterioration in the quality of workers performance brought about by substituting capitalistic social relations to the traditional patron–client relationships between landlords and paysans.

Final considerations

It should be noted, first of all, that the endemic hunger of the Brazilian people is not the consequence of insufficient food production. Hungry people do not represent a market for the agricultural sector if their income is not enough to purchase food with. Sen (1981) shows that even in the great famines in certain regions of the world, food availability was not the cause of the catastrophe. In Bengal in 1943, a critical year, rice production was the highest in all its history. In Ethiopia in 1972–3 food production fell by only 7%. Rice production and its availability per capita in Bangladesh in 1974, the worst period, was the highest in many years. In all these cases people died because they did not have money to buy food. There was an entitlement failure: a direct entitlement failure caused by unemployment, taking from the workers and landless peasants the only thing they had – their work capacity; and an indirect entitlement failure caused by rising food prices (as a result of speculation and other problems in food distribution) compared to wages. Thus in Brazil, if there is not sufficient food produced to feed the people adequately, this is principally because of the low income of the majority of the Brazilian population. It is useless to increase food production without modifying income distribution.

It seems that in Brazil real wages could be raised through a lowering of food prices brought about by the reorganization of the commercial structure. Apart from the problems arising out of a transport system which cannot distribute agricultural production efficiently, a poor storage system, and so on, the *atravessadores* are chiefly responsible for the enormous differential between the prices received by the farmers and those paid by the consumer. As we saw above, this commercial class appeared in the colonial period and has been active since then. The elimination of this class would surely allow an important rise in the prices received by the farmer and at the same time a reduction of the prices paid by the consumer. Thus the increase in agricultural production would be followed by a growth in the market. This process could be strengthened as new agricultural techniques are introduced which raise the productivity per hectare and reduce production costs.

It is difficult to know how much real wages could be raised through this mechanism. It seems reasonable to suppose that, at best, the increase in wages would be enough to provide at least a reasonable diet with, perhaps, some money left over for the purchase of industrial products. But it would be insufficient to give rise to a new pattern of industrial growth. For that, the involvement of the whole society is needed, mainly that of a labouring class organized in free unions. However, it is unlikely that society as a whole and the labouring classes in particular could carry out such a process of income distribution when the mass of unemployed and underemployed is constantly growing because of the rural exodus.

It is difficult to believe that the urban-industrial sector, even with fast rates of growth, could absorb this mass of workers at a time when unemployment is a reality for millions of workers in developed countries. In other words, it does not seem possible for an underdeveloped country, with large numbers of unskilled workers, to accelerate the rate of growth of the urban-industrial sector in order to absorb the structural unemployment, especially when computers and robots are taking over factories and offices. A rural/urban labour ratio like that of the USA (where only around 3% of the active population work in the rural sector) probably cannot be reproduced in any other country in the world without causing serious problems of unemployment. Even for the USA maintaining this situation seems problematic in the long run.

Thus it seems absolutely fundamental, in order to restructure the Brazilian economic model, for the agricultural sector to become an important source of employment. In this context, the existing agrarian structure represents a serious obstacle. It would seem impossible to persuade those holding a monopoly of land in Brazil to abandon extensive cattle breeding, labour-saving mechanization and land disuse in favour of a higher labour absorption. A large contingent of farm workers, with a minimal consciousness of their rights, is directly opposed to the interests of the employers, for whom the right to private property is absolute. For them, as a personal belonging land can be used or not used, conserved or destroyed; as a precious jewel it can be treasured and accumulated, and used as a guarantee for the access to new sources of capital accumulation.

In short, the solution to the problem of food production in Brazil is complex and requires action on three distinct and interrelated levels: the structure of production, the structure of commercialization and the structure of consumption. Concerning the structure of production the problem has two dimensions – one purely technical, and other political – which are not entirely independent. Technically, new ecologically balanced (and therefore more energy-efficient), relatively labour-intensive agricultural systems should be developed; politically, the ideal solutions can be limited by the power of the agro-industrial 'lobby' and of the whole rural oligarchy. It would be necessary to establish as a basis for the legitimacy of private ownership of land its rational use as a source of employment and food production, in ecologically sustainable agrosystems. Concerning the commercial structure, the problem is also twofold; technically, there should be an improvement of the whole infrastructure: roads, silos, efficient transport. This would of necessity narrow the possibilities for speculation with agricultural products, but would not eliminate the speculators, a task which would require repressive political action. Finally, concerning the structure of consumption, the buying power of the great mass of population with a low income should be increased. An increase in real wages could first be brought about by offering more of agricultural products at a lower price, which would result from the above mentioned modifications in the structures of production and commercialization.

Notes

1 Discussions of agricultural policy in the development literature are generally concerned with facilitating the long-term transfer of capital, labour, raw materials and wage goods in favour of the secondary and tertiary sectors of the economy and also with allowing agriculture to be a market to industrial products.

2 *Posseiros* are landless peasants occupying abandoned land belonging to an absentee land owner.

3 Historically this kind of tenancy arrangement was an efficient way of organizing and controlling the work process; it made it possible to avoid problems arising from seasonal characteristics of labour demand in agriculture. The wide availability of land enabled the landlord to transfer to the peasants the costs of producing their own food in the periods of low labour demand in the agricultural production process.

4 It should be noted that the thesis claiming agrarian reform had penetrated even certain factions of the dominant class. The military government installed in 1964 promulgated a new law concerning the land (Estatuto da Terra) that was more advanced and reformist than the precedent promulgated by a government considered as left-wing. Obviously, this new law has never been applied.

5 Wheat is cultivated in rotation with soybeans not because it is good practice in agronomical terms, but because the production process of wheat, like soybeans, can be mechanized in all phases and because it earns a high credit subsidy. For the farmers it pays to grow wheat independently of its productivity (and despite soil degradation in the form of erosion).

6 Big national and multinational groups (Volkswagen, Ford, etc.) established enormous farms in the Amazon region, covering tens of thousands or even hundreds of thousands of hectares, for extensive cattle breeding with a very low productivity (50 kg of meat per hectare per year). One of the methods generally used to prepare the soil is to spray herbicides over the forest to kill it and then to burn it. The weak soils that cover the major part of this region are thus exposed without protection to the characteristic heavy rains, causing violent soil decay. The productivity falls sharply during the first years after the seeding of grass. But this is no problem for these owners since the land costs for

them are nil. The land is paid for by the Brazilian people as a whole, since the money comes through tax exemptions.

7   It is of course in their interest to declare the land as non-arable to avoid land taxes. The existence of unused land is recognized by the government, which has attempted to stimulate its utilization. According to the EMBRAPA's president, those lands are 'fallow'; therefore it is necessary to invest in order to make it productive (Alves 1981).

8   Even in the Japanese case there is evidence that goes against the simplistic hypothesis of Hayami and Ruttan (1971). Ishikawa (1981) observes that during the secondary phase of modernization the labour:land ratio continued to rise, in spite of the increase in the wage:land rent ratio. What should be noted is that, independently of the statistical confirmation of this evidence, the criteria for technological choice of the Japanese farmers were not economical in the sense of optimizing the allocation of factors. They used criteria which were connected to a compromise between social classes, resulting from traditional community relationships which have not only remained, in spite of the modernization of Japanese society, but also become one of its dynamic elements; and not exclusively in agricultural modernization but also in industrial expansion, through the peculiar labour organization of the large Japanese industrial groups.

9   In Java the traditional tenancy arrangements gave rise to what Geertz (1963) called 'shared poverty'.

10   In our opinion, there are two main reasons for the great popularity of the Hayami and Ruttan model: the first is related to the fact that the conception of a 'meta-production function' can solve the problems raised by the neo-classic unrealistic hypothesis of perfect plasticity of capital in the short run (Joan Robinson's 'jelly-like capital'); the second, and more important, is its extreme convenience from an ideological point of view, as it allows the class struggle for possession of land to be reduced to a simple matter of relatively production factor endowments.

References

Alves, Eliseu R. A. 1981. A produtividade da agricultura. *Guia de EMBRAPA* (Brasilia).

Carvalho, F. C. 1978. *Agricultura e questao agraria no pensamento economico Brasileiro (1950/1970)*. Masters thesis, University of Campinas (mimeo).

Castro, A. B. 1975. *Sete ensaios sobre a economia Brasileira*. 2 vols. Rio de Janeiro: Editora Forense.

FIBGE (Fundaçao do Instituto Brasileiro de Geografia e Estatistica). 1981. *Anuario Estatistico*. FIBGE.

Furtado, C. 1963. *Formaçao economica do Brasil*. Brasilia: Ed. Universidade de Brasilia.

Geertz, C. 1963. *Agricultural involution: the process of ecological change in Indonesia*. Berkeley and Los Angeles: University of California Press.

Georgescu-Roegen, N. 1975. Energia y Mitos Economicos. *El Trimestre Economico* **168** (Oct.–Dec.).

Graziano da Silva, J. 1980. *Progresso técnico e relaçoes de trabalho na agricultura paulista.* PhD thesis, University of Campinas (mimeo).

Graziano Neto, F. 1982. *Questro agraria i écologia. Critica da moderna agricultura.* São Paulo: Brasiliense.

Hayami, Y. and V. W. Ruttan 1971. *Agricultural development: an international perspective,* Baltimore: Johns Hopkins University Press.

IPARDES (Instituto Paranaense de Desenvolvimento Economico e Social). 1981. *Boletim de Analise Conjuntural* **3**(1).

Ishikawa, S. 1981. *Essays on technology, employment and institutions in economic development.* Tokyo: Kinokuniya Company.

Linhares, M. Y. and F. C. Teixeira da Silva 1981. *Historia da Agricultura Brasileira. Combates e Controvérsias.* São Paulo: Brasiliense Ed.

Lombardi Neto, F. and J. Bertoni 1975. Tolerancia de Perdas de Terra para solos do Estado di São Paulo. In *Technical Bulletin No. 28.* Campinas: Instituto Agronomico de Campinas.

Mazuchowski, J. Z. 1980. A Experiéncia Brasileira no combate à erosao rural. In *Simposio sobre Controle da Erosao,* Curitiba.

Munhoz, D. G. 1982. *Economia agricola. Agricultura – uma defesa dos subsidios.* Petropoles: Ed. Vozes.

Paiva, R. M., S. Schatten and C. F. T. Freitas 1973. *Setor agricola do Brasil.* Ed. Melhoramentos. São Paulo.

Paschoal, A. 1983a. O onus do modelo da agricultura industrial. *Revista Brasileira de Tecnologia,* CNPq, **14**(1).

Paschoal, A. 1983b. Biocidas – morte a curto e a longo prazo. *Revista Brasileira de Tecnologia,* CNPq, **14**(1).

Pastore, J. and Eliseu R. A. Alves 1975. A reforma do sistema Brasileiro de pesquisa agricola. In *Agricultural Technology and Development,* C. R. Contador (ed.). Rio de Janeiro: Economic and Social Planning Institute/ Research Institute.

Pearse, A. 1980. *Seeds of plenty, seeds of want: social and economic implications of the Green Revolution,* Oxford: Clarendon Press for UNRISD.

Primavesi, A. 1980. *Manejo ecologico do solo. A agricultura em regioes tropicais.* São Paulo: Nobel Ed.

Rangel, I. 1963. *A inflaçao Brasileira.* Rio de Janeiro: Tempo Brasileiro.

Rodale Press 1981. *Empty breadbasket? A study of the US food system by the Cornucopia Project.* Richmond: Rodale Press.

Romeiro, A. R. and F. J. Abrantes 1981. Meio ambiente e modernizaçao agricola. *Revista Brasileira de Geografia* (FIBGE), **43**(1).

Sachs, I. 1980. *Stratégie de l'écodeveloppement.* Paris: Economie et Humanisme et Editions Ouvrières.

Sanders, J. H. and V. W. Ruttan 1978. Biased choice of technology in Brazilian agriculture. In *Induced innovation: technology institutions and development.* H. P. Binswanger and V. W. Ruttan. Baltimore and London: Johns Hopkins University Press.

Schultz, T. W. 1974. Is modern agriculture consistent with a stable environment? In *The Future of Agriculture – Technology, Policies and Adjustment.* Oxford: Agricultural Economics Institute.

Scroccaro, J. L. 1980. Consideraçoes gerais sobre o estudo do transporte de solidos do rio Parana. In *Simposio sobre controle de erasao.* Curitiba.

Sen, A. 1981. *Poverty and famines: an essay on entitlement and deprivation.* Oxford: Clarendon Press.

Sorj, B. 1980. *Estado a classes sociais na agricultura Brasileira.* Rio de Janeiro: Zahar Ed.

Swaminathan, M. S. 1982. *Science and integrated rural development.* New Delhi: Concept Publishing Co.

Tomasini, R., W. Wunsch, and J. Portella. 1981. Uso de energia e manejo racional do solo. In *Anais do II Congresso Brasileiro de Energia.* Rio de Janeiro.

5 Trees: appropriate tools for water and soil management

RATHINDRA NATH ROY

Introduction

Around AD 900 one of the major centres of Maya civilization, in the lowlands of Guatemala, collapsed (with a population decline of 90% in one decade) while at the peak of its agricultural, cultural and architectural development. This event has been analysed, using the latest techniques of paleoecology, and it was shown that the civilization succumbed to cumulative environmental stress: agricultural expansion and population increase led to deforestation and mounting pressure on croplands. These, in turn, led to the loss of topsoil and a gradual decline in the land's productivity, which resulted in water shortages, silting and flooding and culminated in the civilization's loss of its subsistence base (Deevey *et al.* 1979).

Historians traditionally attributed the collapse of civilizations in Mesopotamia to outside invaders. However, recent findings suggest that, because of a lack of drainage, the irrigation system that allowed the civilizations to flourish caused the water table to rise, resulting in waterlogging and salinization of the soil, which in turn destroyed their resource base (Jacobsen & Adams 1958).

In north-west India, exquisite miniature paintings of the Rajasthan School depict scenes of thick and lush forests and the local royalty hunting a variety of wild animals. These paintings depict the situation a few hundred years ago in areas which are now best described as deserts. Fatepur Sikhri, the architecturally extravagant capital city built by the great Mogul Emperor Akbar in the 16th century, was stillborn. It died even while it was being populated because it ran out of life-sustaining water. The peoples of rural India today are facing repeated onslaughts of drought and flooding. Women in several parts of the country have to spend hours every day collecting drinking water, which is often polluted. The inadequate supply and poor quality of water contributes to poor health, disease and death. To the farmer, a shortage of water spells disaster; within his framework, less water for agriculture can mean the end of existence as he knows it. However, the farmer may not understand the close causal relationship between denuding the land of its vegetative cover, soil erosion and a shortage of water.

The rise and fall of civilizations have often been linked to the quantity and quality of topsoil, and therefore to water availability, and have often left deserts in their wake. What separates modern civilization from the Maya is that some of us now understand our predicament and know our present development path to be ecologically unsustainable. However, in common with past civilizations who eroded and thirsted their way out of existence, the reality of our predicament may not be understood by the majority of people who are working the land, or by our leaders, who seem to be too interested in short-term gains to stem the tide of destruction.

The World Bank describes India as a mixed crop zone with a range from warm temperate to arid to humid tropical (Walters 1982). The climate of most areas is characterized by long periods of intense sunshine interspersed with heavy monsoon downpours concentrated over one or two short periods – a most difficult climatic situation for farmers to cope with. In this kind of distinctly seasonal tropical climate, the delicately balanced and well-adapted natural forest seems to represent an ecologically optimal model for agriculture. The trees protect the soil from excessive heat and the violence of the monsoon winds. Moreover, the multi-layered vegetation protects the soil from erosion by runoff and allows rain water to percolate into the ground where it restores aquifers, which may be tapped by the vegetation during the long dry spells. The soils, which are often infertile, are naturally enriched by the rapid decomposition of organic matter and so are able to sustain a varied and highly productive ecosystem. The streams and rivers draining the forested catchments flow perennially with clean water, and simultaneously the groundwater levels remain adequately high.

Forest clearance for agriculture exposes the soil to the baking sun and to the violence of the monsoon. Intense solar radiation causes soil temperatures to rise, increased water loss through evaporation and the destruction of soil organic matter. The monsoon rains crush and compact the exposed soil and thereby inhibit root penetration, they also cause waterlogging and leaching, and finally through erosion they contribute to siltation, which is associated with flooding downstream. Since forest clearance reduces the water-holding capacity of the land and increases storm runoff, it renders the flow of streams and rivers less perennial and reduces aquifer recharge. Poor soil management often leads to a reduction in water availability, which, in rain-fed areas, can only be corrected by improved soil management.

In the past, when population pressure was less, the traditional Indian farmers apparently understood their geoclimatic predicament. Traditional agriculture involved the cultivation of small plots (protected by windbreaks and tree cover), various crop rotation systems and long fallows. Since their demands on the land were comparatively low and their techniques were based on the sound principles of organic husbandry, these farmers were able to maintain some form of stable equilibrium with the natural environment.

Circumstances have changed dramatically since then. The population has grown rapidly and continues to do so. The politico-economic set-up has been transformed through feudalism and colonial domination to independence and a uniquely Indian brand of democracy, which appears to be a manifestation of centre-periphery domination and exploitation, occasionally reinforced by international interests.

Agriculture and soil management have been required to generate food surpluses to support the growing population and satiate the increasing demands of industry. Earlier in this century agricultural output was raised by expanding the area under cultivation; it was then that large-scale deforestation began. In 1950, modern technology was applied to Indian agriculture for the first time. Increased crop yields were achieved through intensive use of energy in the form of hybrid seeds, chemical fertilizers, mechanical tillage, pesticides and large-scale irrigation. The period between 1950 and 1972 was apparently unique with regard to agriculture throughout the world because the amount of food production doubled. However, the total costs associated with the spread of modern agriculture were also impressive, for they included 'scientific' abuse of the land. Such abuse became so severe that 20% to 33% of the world's croplands are now facing water shortages and losing topsoil at a rate that will undermine their long-term productivity (Brown 1982). B. B. Vohra, chairman of India's National Committee for Environmental Planning, estimates that India's yearly contribution to this loss of topsoil is about 6 billion tonnes.

Assuming that a centimetre of soil one hectare of land weighs approximately 138.5 tonnes and that the average thickness of topsoil is 20 cm, 6 billion tonnes is equivalent to approximately 2 million hectares of cultivable land being wasted each year, in India alone (Vohra 1980). The dream of agricultural self-sufficiency and food surplus, reinforced by two decades of increasing crop yields, came to an abrupt end in 1972, when the USSR was forced to purchase wheat at the open world market and famine and drought returned to India and Africa after an absence of a quarter of a century. Since 1972 the gains in agricultural output have barely kept pace with population growth and the rate of agricultural growth shows a clear loss of momentum (Brown 1981).

Mismanagement of soil and therefore of water has created a critical situation, the seriousness of which is difficult to grasp. It has not yet been fully appreciated that water and soil are resources subject to scarcity just like oil and gas. This is partly because of the subtlety of the catastrophe. Harvests are frequently inadequate, but the land continues to produce and the farmers manage to subsist; malnutrition debilitates and often kills infants, but outright starvation remains rare; unemployment and underemployment continue to increase, but small-holding and landless families continue to survive. Moreover, despite the lowering of water tables and the erratic incidence of monsoons there is still enough water for people to drink and for cultivation. These observations may give the impression that there is no crisis, but soil erosion and water shortages, associated with the mismanagement of natural resources, continue apace.

The water and soil crisis in India

The breakdown of land use in India is shown in Table 5.1, but these statistics are misleading, because, of the 83 million hectares classified as forests and permanent pastures, only about 35 million hectares are actually under tree or grass cover. Thus about 88 million hectares of the land under 'biological production' are really unproductive and exposed to erosion. In addition, 87 million hectares of arable land are subject to serious soil erosion and at least 10 million hectares are either waterlogged or saline (Vohra 1980). Therefore, while no definite or 'official' figures are available, it appears that approximately two-thirds of the land which is nominally biologically productive is in a deteriorated condition.

The water-holding capacity of land subjected to soil erosion is drastically reduced and a large proportion of its precipitation input is lost as surface runoff. The consequent reduction in aquifer recharge lowers the water table and makes agriculture difficult, if not impossible, during the long dry seasons. Hence, land use is affected by periodic droughts. Thus soil erosion and deterioration are precursors of the water crisis. In addition to their impacts upon percolation, runoff and recharge, they also affect the quantity and distribution of precipitation by increasing the land's albedo (the

Table 5.1 Land use in India.

Land use	Area (millions of hectares)
total area[a]	305
non-productive (desert)	21
settlements and industry	18
biological production	266
arable	143
forests	75
fallow	23
cultivable waste land	17
permanent pasture	8

Source: Derived from Vohra, 1980.
[a] Total area for which statistics are available.

percentage of incident solar radiation reflected from the land surface) and by reducing evapotranspiration (an important source of atmospheric moisture). Surfaces with high albedos generate thermal currents that affect microclimates and result in either less or wasted precipitation (for example over the sea and over cities) (Charney *et al.*, 1977). Erratic droughts and rain over previously dry areas is becoming increasingly more common in India. It is noteworthy that, whereas in industrialized countries water-use is predominantly industrial and domestic, in India it is more than 90% agricultural. Thus if agriculture faces a water crisis, then so does the overall economy. A large proportion of Indian agriculture is rain-fed and the solution to droughts and flood problems is often seen to be irrigation. Heavy investments have been made in irrigation schemes, and a disproportionately large share of India's agricultural output is produced by irrigated farming. Unfortunately, however, irrigation schemes also depend on river systems fed by precipitation, both of which are affected by denudation and erosion. Moreover, irrigated lands are subject to the new ecological threat of soil destruction due to waterlogging and salinization. Farmers rarely consider an investment in drainage to be worthwhile and tend to overuse water when it is made abundantly available. B. B. Vohra (1980) estimates that at least 10 million hectares of India's irrigated land is now waterlogged and saline and that this has resulted in agricultural productivity losses of 20% or more (Vohra 1980). The reclamation of waterlogged and saline soils takes several years and prodigious amounts of capital – costs which India cannot afford.

Another problem is that India's leaders have chosen the urban–industrial route to development which involves both a rapidly increasing demand for water and the pollution of previously clean sources. The urban–industrial sector has considerable economic and political power, so that it can divert water away from agriculture and thus worsen an already bad situation. With about two-thirds of India's land affected by erosion, water shortages and salinity, and with the added threat of pollution and increasing urban–industrial demand, the country it appears will be facing a catastrophic problem in the 1990s, if not earlier.

Etiology of the water and soil crisis

Soil erosion and water shortages are symptomatic of socio-economic problems which also need to be addressed. It is known that erosion is promoted by exposure of the soil to sun, wind and rain (especially on steep land) and that it may be further aggravated by the structural quality of the soil (loose, crumbly soil without the binding of organic matter is most susceptible); but what role does agriculture play in this process? Monocropping in rows, which leaves large areas of soil totally exposed, and the intensive use of chemical fertilizers and pesticides, which results in a slow elimination of soil organic matter, both encourage erosion. These are agricultural practices introduced by, and adapted from, the industrialized countries. They were designed for temperate climates, with mild and well-distributed precipitation, low insolation and a naturally imposed fallow during winter, but in India they are applied in arid-humid tropical climates with concentrated torrential precipitation, generally high insolation, and no naturally imposed fallow. The persuasiveness of Western (and, therefore, superior) science and technology is such that they have been readily adopted with little consideration of their relevance to the Indian context. This 'intellectual colonialism' (now neocolonialism) persuades Indian scientists to swear by a *modus operandi* which is incongruent with the objective reality of their country.

It is pertinent to question why such technology transfer occurred. Was the transfer merely happenstance, or were there reasons for it? If so, what were they? A historical analysis of social relations and the accompanying agricultural modes shows an interesting correlation between the mode practised and the needs of the ruling classes. Hunter–gatherer and other 'primitive', self-reliant modes are ecologically sound except at excessively high population densities. Sedentary agriculture may produce a surplus, but its market-orientation is a recent phenomenon which apparently developed with the hardening of class structures in society. In India this process

began in the precolonial, feudal period, but was made more concrete, scientific and efficient by the colonial masters who were
concerned with feeding the British market. After independence the
masters changed but the relations remained; the periphery continues
to toil for the centre, within and without. In such a situation,
short-term productivity and profits are all-important. Modern
agronomy guarantees short-term yield gains, but destroys the
long-term sustainability of agricultural production. Thus current
food requirements are satisfied by Western agricultural practice,
even though it is alien to India's ecological circumstances and the
development of her poor farmers.

However, the ruling classes do have to justify their actions.
Fortunately for them there are several arguments which incidentally
demonstrate the populist concerns of the ruling classes. One such
argument goes as follows: the rapidly growing poor and starving
masses must be fed by means of agricultural intensification, because
of the land shortage; to do so, Western science and technology
(including inputs like hybrid seeds, chemical fertilizers, pesticides,
and vast amounts of water), which has been successfully utilized
elsewhere, must be applied. This is a circular argument, which does
not consider aspects of geoclimatic reality, the politics of food and
its equitable distribution, the long-range sustainability of agriculture, or the dependence on other nations for fertilizer and energy
inputs. Thus, in order to feed themselves, Indians have overfished,
overgrazed, overgrown and deforested their land regardless of the
future. In so doing they have begun to destroy their productive
resource base and engaged in the biological equivalent of deficit
financing.

But what about the farmers' attitudes? Surely one would expect
them to worry about the land and generations to come? In India
smallholders and landless labourers are the predominant group
working on the land, although, as one might expect given the
present politico-economic situation, they own only a small part of
it, which is often poor in quality. Caught in the multiple trap of
poverty and indebtedness, with severely limited options, the poor
farmers' first priority is to survive. They naturally try to extract
maximum outputs using minimum inputs, which are all that they
can afford. They cannot afford the luxury of investing in environmental infrastructure neither can they concern themselves with the
long-term sustainability of the soil. Consequently they often
produce a single cash-crop in the wet season and a few free-range
cattle in the dry season. Thus the majority who work the land are in
situations which lead them to promote environmental destruction
and water shortages through their efforts to survive. The social,

political and economic forces act in a way which gives them no other option.

To summarize, India's water scarcity and soil erosion problems stem from intensive cropping using unsuitable imported agricultural technology, and from the country's socio-politico-economic reality which has mortgaged long-term sustainability and development for short-term profit involving exploitation of the masses who work on the land.

Tree-based ecofarming: a technical solution

This water–soil problem cannot be solved by means of a 'technological fix'. However, a solution to the underlying social, political and economic aspects would be incomplete without an ecologically sound technology, which would allow people to work the land and, at the same time, to conserve soil and water resources. In this section a technical solution is proposed with such a contingency in mind.

An agricultural technology is needed which will protect the soil and enable it to absorb precipitation, allow for perennial surface runoff and also provide the food, fibre, fuel and other materials needed to sustain the population, plus a reasonable surplus. In addition, the technology should be economically viable at the small-farm level.

By listing the inverse of the negative aspects of India's existing agricultural technology, we can provide a brief for the agricultural technology necessary to conserve both water and soil resources. The 'new' agricultural technology would:

(1) protect the soil from the direct impact of rain and maintain the soil structure necessary for it to absorb and hold water, thus reducing storm runoff and facilitating percolation and aquifer recharge;
(2) protect the soil from direct solar radiation, thereby preventing the soil temperature increases which destroy organic matter, reducing soil moisture loss due to evaporation, and reducing weed growth;
(3) reduce wind erosion by the provision of windbreaks;
(4) increase transpiration so as to increase precipitation and decrease surface albedo, thereby preventing disturbances of microclimate;
(5) involve little or no tillage;
(6) utilize poor quality soils;

(7) utilize small quantities of water, and not depend on a continuous water supply;
(8) provide employment throughout the year instead of seasonally;
(9) produce sustainable yields of food, fodder, fuel, fibre, construction material and livestock;
(10) produce sufficient yields on each smallholding to sustain a
family and generate a surplus.

A remarkable technology which has all of these effects has long
existed as the agronomical analogue of the tropical forest. It is a
multi-tier, organic farming system that is based on the cultivation of
trees. This concept has been well developed by Douglas and Hart
(1978) and Howard (1943). All of Howard's research was undertaken in India, where he served as an agricultural officer during the
early part of this century. Until recently this type of work has been
largely ignored, which is a measure of the value-free, open-minded
objectivity of Western science and technology.

However, much more research is needed to help the farmers to
adopt ecologically sound agricultural practices. For example, the
reclamation of poor and deteriorated soils and the increasing of crop
yields to the high levels that are claimed by organic farmers
elsewhere may take several years. In the meantime, problems
associated with lower than average yields will have to be resolved.
Moreover, since new, tree-based agriculture takes a comparatively
long time to produce regular yields, phased introduction schemes
and integrated cropping patterns should be devised and tested to
ensure that farmers will have reasonable and regular outputs prior to
tree maturation. The politico-economic aspects will also have to be
studied carefully and checked in advance.

The difficult process of land reclamation requires the development of water resources to sustain the young trees through the first
few dry seasons. Also, using labour-intensive methods, earthwork
bunds, check dams, contour bunding and gulley plugs have to be
built prior to the monsoon. This would allow runoff to be checked,
rainwater to be collected and stored, soil moisture to be increased
and the groundwater level to be raised. Drought-resistant shrubs
and plants should be grown to provide ground cover, green manure
and compost, and seedlings should be protected (by soil stabilization) to avoid siltation and removal in runoff. Another difficult
problem would be the protection of the plants from animals and
people. This could be achieved by using thorn bushes, baskets,
fencing (where available) and especially by personally guarding the
plants.

The initial phase would be a holding operation, involving restoration of the land–water system while the sustainable form of agro-forestry is developed. Few production gains will be possible during this phase, but the alternatives for providing a sustainable system are extremely limited. Theoretically, the idea of agroforestry is feasible and the results of practical tests are encouraging enough to justify optimism, further research and implementation (Auroville Communications Group 1980).

The economic viability of agroforestry

The concept of economic viability will first be considered in terms of the framework within which agroforestry should be viewed. The primary problem with modern agriculture is the imperative to produce vast surpluses so that it can effectively contribute to the wider economy. The generation of surpluses, beyond survival needs, is ecologically unnatural and always requires inputs of higher quality energy to supplement those of the relatively inefficient natural systems. Therefore, productivity evaluations of agroforestry, which is analogous to natural tropical forest, and comparisons with Western agriculture should be made in energy efficiency terms and not in terms of production outputs alone. Western agriculture appears to produce greatly, but in doing so it utilizes large amounts of energy which are rarely taken into account.

Ecologically sound agroforestry will be sustainable, but may not be able to satisfy the needs of an economy which is based on energy intensive lifestyles. It would be thermodynamically inappropriate to attempt to satisfy current (and future) high energy and environmentally destructive demands through alternative, ecologically sound modes of production. This would suggest that the needs (as different from wants) of the Indian people would have to be modified on the basis of new lifestyles and ethics such as conservation, voluntary simplicity and concern for wellbeing of others and future generations. It would also be necessary to consider the philosophical stance and the socio-politico-economic and ethical framework of the nation's leadership which would motivate the leadership into giving ecological sustainability, harmony with nature and 'true' development higher priorities than exploitation of nature to satisfy short-term greed.

A second level of analysis would be to consider viability with regard to the development of social financing that would allow small farmers to sustain their families through the long maturation period. This idea has previously been pursued in housing programmes, medical care and irrigation projects. In most cases, the returns on investment in such schemes are far less concrete and beneficial than they would be in ecologically sound agroforestry. It would be worthwhile to determine the cost/benefit balance of such a system in which costly inputs like water, fertilizer and energy are reduced while long-term sustainability at high output levels and soil and water conservation are ensured. Moreover, the savings on flood control and flood damage would surely make the investment in ecologically sound agroforestry an economically attractive option.

Sociopolitical impediments to implementation

The mere existence of the solution to a problem, which is technically and economically feasible, is not sufficient to ensure its promotion and implementation, because of the social and political impediments to such a programme.

At the personal level it means that millions of farmers would have to change their ways of husbanding soil and water resources and some of their traditionally accumulated knowledge might have to be discarded, modified and supplemented. In the Indian context, where agricultural knowledge has become unchanging and stagnant and is often communicated out of context (to be learnt by rote and applied blindly), this task seems almost impossible to achieve. However, the breakdown of traditional social structures, the alienation of the youth and the associated 'generation gap' could actually become positive factors in overcoming such mental inertia and help in the promotion of the new agroforestry technology.

Since water shortage is a readily visible and understandable phenomenon, as compared with the exacerbating problem of soil erosion, it should be tackled first. It is important to begin any action with the problems and needs that the people perceive as important, and then to use the felt needs as a lever to promote a critical understanding of the environment, which, in turn, would surface the actual needs and problems.

Agriculture needs land, but, in India, this resource is not freely available; it is either privately owned, or controlled, as in the case of waste land, by the government and the local community. For the land reclamation envisaged to be undertaken, principally by smallholders and landless farmers, a redistribution of land either through lease or freeholdings would be essential. This implies a confrontation with the dominant land-owning class, which has successfully resisted all but nominal land reform in India. Land is a form of wealth and power, which will not be easily given over in the interests of conservation and true human development, it will have to be taken.

At higher levels of social aggregation, once landless farmers are given access to land for self-reliant agriculture, they may choose not to remain on the lower rungs of Indian society and a dramatic, violent reshuffling of rural power structures may occur. Landless peasants and smallholders are presently the primary sources of cheap labour, which is used to profit the larger farmers. When they are given access to more land of their own, who will labour on the larger farms and what will become of the larger farmers and absentee landlords?

Another question to be considered is that, although ecologically sound agroforestry focuses on soil and water conservation, as opposed to feeding the cravings of a runaway economy, farmers will still have to sell their surpluses. Should this be done within a framework of local self-sufficiency or will the larger markets siphon off local products so as to create a cash-crop economy as it did with the milk and egg production schemes? The present political system, which is supported by the rural rich and the urban industrial economy, will have to be reorganized to accommodate the new power configurations.

These and other considerations indicate why the change to ecologically sound agroforestry is likely to be blocked, especially by the powerful and by those who have something to gain by maintaining the status quo. However, there are grounds for optimism as the severity of India's water and soil crisis and the limitations on future resource-use options are increasingly recognized. These are not problems which the Indian people are likely to accept with resignation.

But who is going to take the necessary action to solve the water and soil crisis and who will be able to overcome the opposition and make the new agriculture a reality? It is not reasonable to expect those presently in power to restructure the system so as to make themselves less powerful. Nor is it reasonable to expect government, with its diffuse responsibility, lack of accountability and innate support of the powerful, to bring about the necessary change. However, the rural masses, who stand to gain greatly from land reform and the new agriculture, could overcome the opposition by force of numbers. Concerned scientists, who claim to know and to have solidarity with the interests of the people and with nature conservation, are privileged to be able to help people to do what needs to be done.

The urgent needs are to test the idea and simultaneously inform, persuade, lobby, motivate and mobilize people into critical consideration of the existing situation and the available alternatives. Subsequently it will be necessary to organize social, political, economic and government forces in directions that will enable and encourage people to realize their interests while working towards the desirable social and ecological goals.

Pathways to action

Given the likely opposition to the new agriculture, as a solution to the soil and water crisis, a strong popular movement will be required to overcome it. Several pathways to action may be necessary to deal with the various aspects of this problem and to encourage communications amongst the groups involved.

Before recommending such pathways it is necessary to anticipate a pre-emptive strike from those opposed to the new agriculture. Often, the best means to destroy an idea to which one is opposed is to pretend to pursue it and simultaneously ensure that its goals are not achieved. The various Social Forestry Schemes currently being implemented throughout India, with generous aid from international agencies and foreign governments, are a topical example. These apparently impressive schemes propose that waste lands and fallows should be planted, primarily by the Forest Department, with economically useful trees. The supposed goals are to provide rural areas, which have scarcities of fuel and fodder, with regular supplies of both, and simultaneously to improve the environmental situation. However, in practice most of the new trees had potential end-uses in industry. An implicit assumption in the Social Forestry Schemes has been that if enough forest products are produced they will naturally be used by all. This is at best naive, for national market forces are far stronger than local demand. It has resulted in a massive flow of forest products, which were meant to solve local problems, to the urban–industrial areas.

In pursuit of quick and large profits, several big farmers grew forests instead of food crops on high quality agricultural land. Several studies have taken a critical view of the Social Forestry Schemes and have pointed out the essential futility of the concept, which if developed with the environment and people in mind could have been highly convivial (Alvares 1981).

Three overlapping and interdependent pathways are recommended to help develop a movement for ecologically sound agroforestry to solve the water and soil crisis. These will have to be built upon in the light of practical experience. First, there is a need to communicate with people, persuade them of the idea and mobilize them into activity. The most important group to be communicated with are those most involved; the smallholders, the farmers of marginal land, the landless labourers and those who live and work on forest land. This group will have to be made aware of their environmental problems so that they can recognize their predicament and the options available to them. In short, they will have to create the means to communicate knowledge amongst themselves. The general public, especially the middle class, have to understand how the environment affects their lives. The mass media, schools and colleges, talks to clubs and associations and various other means could be used to communicate with this group. Finally, the opinion-makers, legislators, political and scientific groups and international agencies will have to be provided with information and legislative support and influenced by pressure groups.

Secondly, there is the research pathway in which information and knowledge will have to be generated to enable the practical implementation of ecologically sound agroforestry programmes at the field level. Small centres should be established in different land types to undertake research into organic farming techniques, soil and water conservation, forest farming and appropriate integrated farming in order to provide feasibility studies, detailed costings, regional manuals, training and demonstrations.

Thirdly, and perhaps most difficult, would be the pathway of field action. Non-governmental organizations, voluntary groups, political groups and people's organizations must work to resolve the problems associated with land ownership, land reform, leasing community and government waste lands, the financial aspects of the new agriculture, field demonstrations, the establishment of information centres, the mobilization of support and the fight to overcome opposition and make leaders accountable to the people.

The pathways considered above are necessarily vague for it is impossible to produce a handbook for political and social action. The ideas and means of implementation will have to be tested and tried at the field level and then put into practice. The task is enormous, but the temptation to create one large agency to administer it should be avoided. Instead action should be decentralized and thereby more consistent with the overall concept.

By way of conclusion, it should be remembered that what separates contemporary civilizations from those which have collapsed in the past, due to soil and water crises, is that today we understand our predicament and know that our development path is not sustainable. It remains to be seen whether we are going to apply this knowledge effectively.

RATHINDRA N. ROY   125

References

Alvares, C. 1981. The social forestry con game. *Deccan Herald* (Bangalore), 12 November.

Auroville Communications Group 1980. Bringing the land back to life: environmental regeneration in India. *Soil Assoc. Q. Rev.* **11**(3), 3–4.

Brown, L. R. 1981. World population growth, soil erosion, and food security. *Science* **214,** 995–1002.

Brown, L. R. 1982. R & D for a sustainable society. *Am. Sci.* **70,** 14–17.

Charney, J. G., W. J. Quick, Shu-Hsïen Chow and Jack Kornfield 1977. A comparative study of the effect of albedo change on drought in semi-arid regions. *J. Atmos. Sci.* **34,** 1366–89.

Deevey, E. S., D. S. Rice, P. M. Rice, H. H. Vaughan, M. Brenner and M. S. Flannery 1979. Mayan urbanism: impact on a tropical karst environment. *Science* **206,** 298–306.

Douglas, J. S. and R. A. de J. Hart 1978. *Forest farming: towards a solution to problems of world hunger and conservation.* Emmaus, PA: Rodale Press.

Howard, A. 1943. *An agricultural testament.* New York and London: Oxford University Press.

Jacobsen, T. and R. M. Adams 1958. Salt and silt in Ancient Mesopotamian agriculture. *Science* **128,** 1251–8.

Vohra, B. B. 1980. *A policy for land and water.* Sardar Patel Memorial Lecture, New Delhi.

Walters, H. E. 1982. Agriculture and development. *Finance and Development* **19**(3), 6–11.

White, R. R. 1977. Resources and needs: assessments of the world water situation. UN Water Conference.

6 Low-energy farming systems in Nigeria

BERNHARD GLAESER and
KEVIN D. PHILLIPS-HOWARD

Introduction

Rural development, including improvements in agricultural production and the wellbeing of rural communities, is a goal shared by many countries. The precise aims and means of achieving such goals vary, but in all cases the transition involves manipulation of land, capital, labour and energy. The last of these has only recently been considered a significant factor in this context.

The supply and use of energy has become a crucial issue in both industrialized and developing countries. This is particularly true with respect to food production. In this chapter we focus on energy use in agriculture, through an analysis of input–output budgets. In so doing, we examine the relationship between agricultural technology and the efficiency of energy use.

Technological alternatives to energy-intensive, industrial agriculture are also considered. These include various energy-saving types which are collectively known as 'ecological farming'. They have the additional advantage of being environmentally sound. It is argued that such alternatives are worthy of greater attention. They may be particularly appropriate for developing countries for both economic and environmental reasons. Indeed, many developing countries already have some form of traditional agriculture which is both energy-saving and environmentally sound. A case study from south-east Nigeria is presented. It describes energy use associated with mixed cropping, particularly the production and protection aspects of agricultural technology.

The concept of technology used here refers to the application of knowledge (*lógos*) by techniques (*téchne*) in the broadest sense. Agricultural technology, as we understand it, is a complete set of techniques by which ecological knowledge is applied to produce food. It is regarded to be interrelated with the sociocultural, economic and biophysical environments of its users.

Energy use in agriculture

Industrialized countries

In industrialized countries, agriculture accounts for 1.5% to 3% of the total commercial energy consumption. The figure may rise to 8% when the energy requirements of food processing and distribution are included (Hauser 1975, Keller 1977, Studer 1978). This may not seem large, but when extrapolated in absolute terms, to the world's population, it amounts to 20×10^{15} kilocalories. This is roughly equivalent to the total conversion of thermal energy throughout the world (Hampicke 1977, p. 36). Evidently, industrial agriculture requires enormous energy inputs in order to feed people.

The magnitude of these inputs, as compared with the outputs in food, can be calculated through 'energy accounting'. The energy rate (ER), the ratio of outputs to inputs, is particularly useful for this purpose. Consider, for example, the case of maize (corn) production in the United States. Between 1945 and 1970 the food energy yield doubled, whereas the energy additions in nitrogenous fertilizers increased by sixteenfold and those in pesticides rose from zero to 22 000 kilocalories per acre. Table 6.1 shows that during the same period, the ER dwindled from 3.7 to a less efficient 2.8 (Pimentel *et al.* 1973).

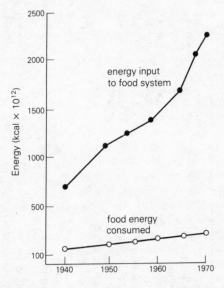

Figure 6.1 Energy use in the food system, 1940–70, compared to the caloric content of food consumed (from Steinhart & Steinhart 1974).

Table 6.1 Energy inputs (kilocalories) in United States corn production.

Input	1945	1950	1954	1959	1964	1970
labour	12 500	9 800	9 300	7 600	6 000	4 900
machinery	180 000	250 000	300 000	350 000	420 000	420 000
gasoline	543 400	615 800	688 300	724 500	760 700	797 000
nitrogen	58 800	126 000	226 800	344 400	487 200	940 800
phosphorus	10 600	15 200	18 200	24 300	27 400	47 100
potassium	5 200	10 500	50 400	60 400	68 000	68 000
seeds for planting	34 000	40 400	18 900	36 500	30 400	63 000
irrigation	19 000	23 000	27 000	31 000	34 000	34 000
insecticides	0	1 100	3 300	7 700	11 000	11 000
herbicides	0	600	1 100	2 800	4 200	11 000
drying	10 000	30 000	60 000	100 000	120 000	120 000
electricity	32 000	54 000	100 000	140 000	203 000	310 000
transportation	20 000	30 000	45 000	60 000	70 000	70 000
total inputs	925 500	1 206 400	1 548 300	1 889 200	2 241 900	2 896 800
corn yield (output)	3 427 200	3 830 400	4 132 800	5 443 200	6 854 400	8 164 800
kcal return/input kcal	3.70	3.18	2.67	2.88	3.06	2.82

Source: Pimentel *et al.* 1973.

Table 8.2 Energy use in the United States food system, 1940-70 (all values are multiplied by 10¹² Kcal).

Component	1940	1947	1950	1954	1958	1960	1964	1968	1970
Agriculture									
fuel (direct use)	70.0	136.0	158.0	172.8	179.0	188.0	213.9	226.0	232.0
electricity	0.7	32.0	32.9	40.0	44.0	46.1	50.0	57.3	63.8
fertilizer	12.4	19.5	24.0	30.6	32.2	41.0	60.0	87.0	94.0
agricultural steel	1.6	2.0	2.7	2.5	2.0	1.7	2.5	2.4	2.0
farm machinery	9.0	34.7	30.0	29.5	50.2	52.0	60.0	75.0	80.0
tractors	12.8	25.0	30.8	23.6	16.4	11.8	20.0	20.5	19.3
irrigation	18.0	22.8	25.0	29.6	32.5	33.3	34.1	34.8	35.0
Subtotal	124.5	272.0	303.4	328.6	356.3	373.9	440.5	503.0	526.1
Processing industry									
food processing industry	147.0	177.5	192.0	211.5	212.6	224.0	249.0	295.0	308.0
food processing machinery	0.7	5.7	5.0	4.9	4.9	5.0	6.0	6.0	6.0
paper packaging	8.5	14.8	17.0	20.0	26.0	28.0	31.0	35.7	38.0
gas containers	14.0	25.7	26.0	27.0	30.2	31.0	34.0	41.9	47.0
steel cans and aluminium	38.0	55.8	62.0	73.7	85.4	86.0	91.0	112.2	122.0
transport (fuel)	49.6	86.1	102.0	122.3	140.2	153.3	184.0	226.6	246.9
trucks and trailers (manufacture)	28.0	42.0	49.5	47.0	43.0	44.2	61.0	70.2	74.0
Subtotal	285.8	407.6	453.5	506.4	542.3	571.5	656.0	787.6	841.9
Commercial and home									
commercial refrigeration and cooking	121.0	141.0	150.0	161.0	176.0	186.2	209.0	241.0	263.0
refrigeration machinery (home and commercial)	10.0	24.0	25.0	27.5	29.4	32.0	40.0	56.0	61.0
home refrigeration and cooking	144.2	184.0	202.3	228.0	257.0	276.6	345.0	433.9	480.0
Subtotal	275.2	349.0	377.3	416.5	462.4	494.8	594.0	730.9	804.0
Grand total	685.5	1028.6	1134.2	1251.5	1361.0	1440.2	1690.5	2021.5	2172.0

Source: Steinhart & Steinhart 1974.

Mechanization, fuel use and fertilizer applications account for most of the energy consumed by industrial agriculture. For example, in 1970, they amounted to 79% of the energy input to US corn production (Pimentel *et al*. 1973) and 81% of that to the US food system as a whole (see Table 6.2). Between 1940 and 1970, total energy use in the US food system increased from 685×10^{12} to 2172×10^{12} kilocalories, whereas the increase in food consumption per capita was so small that the food–consumption curve (see Fig. 6.1) is a reflection of population growth. As a result, the energy ratio decreased from 1:5 (0.2) to 1:10 (0.1). In other words, by 1970, ten energy units were used – in production, processing, distribution, storage and cooking – to provide one energy unit on the dinner table (Steinhart & Steinhart 1974).

The problems associated with such enormous energy use include soaring prices, environmental pollution, resource exhaustion, international dependence and uncertainty and underdevelopment of non-industrial countries. In short, these amount to 'maldevelopment' (Sachs 1979). Clearly, there is a strong case for the examination of alternatives.

A technological alternative to energy-intensive agriculture

The technological alternative is an energy-saving, environmentally sound food system which produces good average yields. Such an alternative exists in the form of 'organic' or 'ecological' farming. It is characterized by the use of naturally converted solar energy, manipulation of nutrient cycles, maintenance of crop diversity, 'biological' methods of pest, weed and erosion control, and the exploitation of the ecological succession. There is little doubt that this alternative can produce high-quality food, with minimal damage to the environment. The question is, can it produce sufficient quantities of food to meet the demands?

A number of comparative studies have shown that the yields, using ecological farming technology, are not equivalent to the highest achieved by industrial agriculture, but that they can reach good average levels. For example, in a three-year investigation of nine ecological farming units in Baden-Württemberg (southern Germany), it was found (Auswertung 1977) that average yields for the region were generally matched or nearly matched. In some cases, the average yields of industrial farms were actually exceeded. For example, a survey in the Netherlands found that ecological farming units produced higher than average yields for seven out of nine crops. These included wheat, barley, oats, beetroot and beans (Ronnenberg 1972, quoted in Vogtmann 1977).

A comparison of 16 'organic' farms and 16 'conventional' farms in the US 'corn belt' revealed little difference in yields. In this case, it was noted that the energy requirements of the organic farms were not more than a third of their conventional, capital-intensive competitors (Lockeretz *et al.* 1975).

In view of these findings, it is not surprising that energy saving in agriculture is considered both feasible and worthwhile (Pimentel *et al.* 1974, p. 464). Concrete measures have been identified which reduce energy consumption in conventional agriculture without a complete change of the farming system (Steinhart & Steinhart, 1974). These measures include:

(1) substitution of chemical fertilizer with organic manure (energy saving $= 4 \times 10^5$ kilocalories per hectare);
(2) partial substitution of chemical with mechanical pest control (10% saving);
(3) substitution of high-yielding varieties, especially in tropical areas, with more resistant varieties.

Developing countries

Agricultural energy consumption in developing countries has traditionally been lower than in industrialized countries (Biswas & Biswas 1975). However, the so-called 'Green Revolution' has partly changed this situation through its promotion of capital and energy-intensive technology. For example, the energy rate of rice and vegetable gardening in Hong Kong during the 1930s was 24.4. By 1971, after the introduction of energy-intensive technology, the ER had declined to an average value of 1.3, whereas for intensively cultivated fields the value dropped as low as 0.13 (Newcombe 1975). Similar examples have been reported from Latin America (Leach 1976), Africa and India (Makhijani & Poole 1975). They clearly demonstrate that agricultural development through the adoption of capital-intensive technology implies massive increases in energy inputs.

Nevertheless, under present economic and political circumstances, the prospects for Third World rural development based on increased energy consumption appear to be severely limited. Many countries must seek improved productivity and higher standards of living while continuing to depend largely on traditional energy sources (Morgan 1980). This is true even for petroleum-rich Nigeria because of problems associated with inflation, population growth, urbanization, distribution, and so on, and the priority given to urban-industrial development.

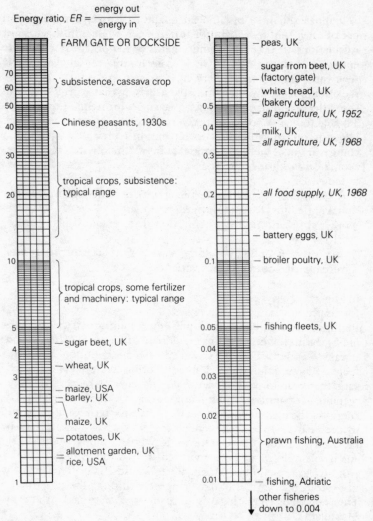

Figure 6.2 Energy rates (ER) in food production (from Leach 1976).

Figure 6.2 shows that the energy rates of traditional agriculture in developing countries vary from 10.0 to 70.0, as compared with 0.1 to 5.0 for agriculture in industrial countries. Evidently traditional agriculture results in food energy outputs which are extremely high compared with the energy used in their production. The energy rate is not the only 'yardstick' for evaluating agricultural production, but it does indicate the relative energy efficiency of different types of agriculture.

It appears that the traditional systems make much more efficient use of their limited 'anthropogenic' energy inputs. However, these calculations are based on edible energy returns for labour and other energy put in (Leach 1976, p. 7). They do not include all the renewable energy inputs derived from natural conversion of solar radiation. Hence, they are only partial energy budgets, but they do show the extent to which traditional agriculture is based upon, and adapted to, use of renewable (solar) energy. An important question here for rural development is how, with limited anthropogenic energy inputs, this type of agriculture can be upgraded using ecological technology which has produced such favourable results in industrialized countries.

A preliminary description of the energy situation in one traditional agricultural system, which could serve as a basis for further work, is presented in the next section.

The case of south-east Nigeria

In Nigeria as elsewhere, attention has been drawn to the possibilities for improving the use of renewable energy resources (Ojo 1978, Phillips-Howard 1978). Some preliminary studies of fuelwood-use in Nigeria have been made (e.g. Morgan 1978, Ay 1980) and, under the auspices of the United Nations University, a project focusing on fuelwood (which also includes wind, solar energy, biogas and fossil fuels) has been initiated (Ojo 1980).

In this section attention is drawn to food, and its importance as an energy-source in the rural Third World. This is done with reference to the results of a recent survey in south-east Nigeria (Phillips-Howard 1980a).[1]

Energy needs and the role of food

The principal energy needs of a rural community, as identified by Makhijani (1976), are as follows:

(1) Agricultural fuels (irrigation, draught power, fertilizers, manufacture of implements, crop processing, food transport, food storage);
(2) energy for cooking;
(3) energy for providing clean domestic water supply which, in some places, includes energy for boiling drinking water;
(4) house heating and warm water for bathing in cold climates;
(5) hot water and soap for washing clothes;

(6) energy for lighting (household or community);
(7) energy for personal transport;
(8) energy for processing and fabricating materials needed for house construction, pots and pans, clothes, tools, bicycles, etc.;
(9) energy for transport of goods;
(10) energy needed to run local health services, schools, government and other community uses.

The energy needs of rural south-east Nigeria (see Fig. 6.3) differ in a number of ways. Agricultural fuels (in the conventional sense of oil, and so on) are little used, since most agriculture in this area is not mechanized. Irrigation is unnecessary for most of the year because of the humid climate, short dry season (two to four months) and the limited period of soil-water deficit. Moreover, the need for irrigation is largely avoided because the annual crops are not planted until 'after the first rains' (January–February) at the end of the dry season. It is limited mainly to the construction of shallow ditches, using hoes, to direct rainwater around the crops. However, along the distributaries of the Niger (e.g. the Sangana, Forcados, Nun and Bonny Rivers) it is common for farmers to dig deeper channels to guide floodwater on to adjacent land.

Tractors are little used for cultivation in south-east Nigeria, and

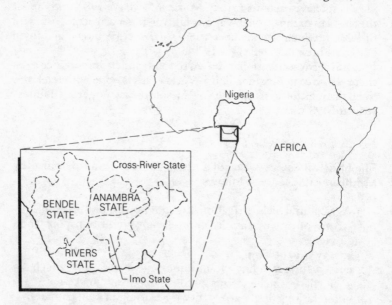

Figure 6.3 The location of south-east Nigeria.

draught animals, such as oxen, are absent – probably due to the presence of tsetse flies (the vectors of trypanosomiasis). Hence, with the exception of mechanized clearance schemes sponsored by government, cultivation is done by hoe. Artificial fertilizers are not usually applied. In most areas, nutrient supply continues to depend on the bush-fallow system. Again, exceptions can be cited. In the areas subject to flooding, annual nutrient replenishment, in the form of silt, permits continuous cropping. Crops (particularly leaf vegetables) are also grown every year at the backs of houses where they benefit from domestic wastes. But these systems do not depend on exogenous energy sources.

Since most of the agricultural tools (including machetes, knives, axes and hoes) are not manufactured in the rural areas, they do represent a small input of exogenous energy. Food processing is undertaken, by women and girls, using simple equipment such as knives and the mortar and pestle (Oke 1979). In towns, and a few villages, mechanical equipment may be available for grinding cassava and maize.

The transportation of agricultural produce is done mainly by head-porterage. However, use of bicycles is also common. In this case, although motive power is provided by human muscular energy, the bicycles themselves are a form of exogenous energy. Goods intended for sale are taken to market by these means or by motorized transport, only the latter involving a substantial addition of fossil fuel.

In the absence of refrigeration, food preservation is difficult in the hot, humid conditions of south-east Nigeria. As Imevbore (1973) notes, if food is not prepared quickly for storage it will rot and have to be thrown away. Much food is eaten soon after it is harvested, so reducing the preservation problem. Nevertheless, Table 6.3 shows that there are a variety of traditional means for storage and preservation of food. They differ according to the type of produce, but, in all cases, they involve the creation, through human labour, of suitable conditions from locally available materials. Such methods constitute a low-energy strategy for reducing the loss of potential (food) energy inputs.

Energy for cooking is mainly derived from fuelwood taken from the farms and the forest.[2] It is the one important energy need which cannot be met through work alone, although human energy is needed to cut, gather and transport the wood.

The domestic water supply of the majority of villages, in which piped water is not available (Olatunbosun 1975), depends on collection, mainly by children, from streams (in other words, human labour). In the few cases where piped water is available,

Table 6.3 Examples of traditional food preservation and storage in south-east Nigeria.

Food type	Method								
	Sun drying	Under shade	In sacks	Under banana leaves	In the soil	In barns tied to racks	Under the roof rafters	In covered baskets	In clay pots
cassava					×				
yam					×	×			×
cocoyam				×					
maize			×				×		
melon	×		×						
pepper	×								
plantain		×							
coconut							×		
pumpkin								×	
tomato			×				×		
palm oil kernels			×						
cacao				×					

Source: Phillips-Howard 1980a.

collection from the communal tap is still necessary. However, the pipe system itself represents a large infusion of exogenous energy. House-heating and hot water for bathing and washing clothes are, for climatic reasons, not necessary. Clothes washing is invariably done by hand, although the soaps and powders which are used in this process comprise a small exogenous energy supplement.

The need for lighting remains, for most people, largely unsatisfied. Those who have it mainly use candles or kerosene lamps. A few wealthy people have diesel-fuelled generators which supply electric lighting.

Energy needs for personal transport, in rural south-east Nigeria, are mostly met by 'trekking' (walking). Exogenous energy in the form of bicycles and motorized transport has, however, increased the mobility of both people and goods.

Traditional house-construction involves use of local materials (wood, leaves, mud, raffia, etc.) and human labour. The most common houses are rectangular, with mud walls, constructed on wooden frames and roofed with thatch. However, there are many concrete houses with corrugated-iron roofs. These are constructed by labour-intensive methods, but the materials incorporate large injections of fossil fuels. Material possessions, including pots, pans and clothes are generally limited to a few basic items. These are often exogenous, even in the remotest rural areas, and represent a small input of energy.

Finally, although the application of traditional medicine (and other endogenous knowledge) continues to thrive in south-east Nigeria (Phillips-Howard 1980b), the provision of local government services is proceeding. All of these, and particularly the health clinics (Madon et al. 1979), require a considerable energy input. However, in this case, as with the other oil-based inputs, most of the services are highly localized.[3] Hence the flow of exogenous energy is concentrated in LGA (Local Government Area) centres and dilute elsewhere.

Evidently, the majority of people in rural south-east Nigeria continue to depend, for most of their energy needs, upon locally available human and environmental resources. With the notable exception of fuel for cooking, they continue to be satisfied through human labour. Hence, it is clear that the energy sources of human activity, that is 'food supplies', play a vital role in the rural energy system. Indeed, as Mascarenhas (1980, p. 83) observed, the problem of food supplies is 'fundamental to the understanding of the relation between man and his environment in the developing world'. The continuity of food-energy inputs is, in contrast to fuelwood, biogas, hydropower, and so on, literally 'a matter of life and death' to subsistence farmers. Food is the one energy source for which they can find no substitute. Therefore, we suggest that human energy use, especially in food production, is worthy of much greater attention, even though the quantities involved may be comparatively small (e.g. Reddy 1979).

In the following section, a simple energy-budget model which focuses on food and its production is presented.

An energy budget for subsistence agriculture

We observed earlier that Leach (1976) calculated the ER of traditional agriculture as the edible energy returns for work put in. In the present context, it is helpful to reverse this calculation, that is, to regard labour as an output which depends upon food as an input.

It is vital for subsistence farmers that their total energy output (TEO) does not exceed their total energy input (TEI). The major elements which comprise these totals are as follows:

$$\text{energy outputs: TEO} = \text{BMO} + \text{BGO} + \text{FAO} + \text{NAO}$$

where BMO = basic metabolic output; BGO = bodily growth output; FAO = farming activity output; and NAO = non-farming activity output.

$$\text{energy inputs: TEI} = \text{FEI} + \text{FWI} + \text{EEI}$$

where FEI = food energy input; FWI = fuelwood input; and EEI = exogenous energy input (tools, fertilizers, pesticides, government services, and so on).

The calculation of such a budget would reveal quantitatively the way and extent to which rural energy needs are satisfied. It would also draw attention to inefficiencies and possible energy-use alternatives. Unfortunately, we cannot yet compute this budget for farmers in south-east Nigeria. However, it is possible to make preliminary comments on the energy-use management. In so doing, we will concentrate on the use of technology, and other means, in food production; that is, the manipulation of farming activity outputs (FAO) to ensure food energy inputs (FEI).

The use of agricultural technology

As noted earlier, the tools used by farmers in south-east Nigeria are very simple. Their use is primarily geared to local subsistence, but also to the cultivation of 'cash-crops'. In broad energetic terms, the available tools are employed, as efficiently as possible, to aid

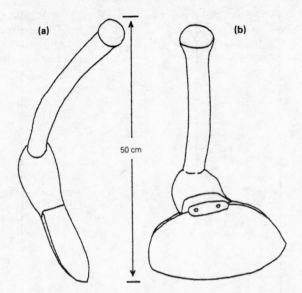

Figure 6.4   The two types of hoe used in south-east Nigeria: (a) narrow-bladed hoe; (b) wide-bladed hoe.

the conversion of incoming solar radiation into edible chemical energy.

For example, consider the use of hoes. Two types are regularly used in south-east Nigeria: the narrow-bladed hoe and the wide-bladed hoe (see Fig. 6.4). The narrow-bladed ones require the expenditure of relatively little energy and permit a farm to be dug comparatively quickly. In contrast, the wide-bladed hoes demand more energy expenditure, and more time, to dig the same farm. However, the farmers know that the wide-bladed hoes destroy the roots of weeds much more effectively. Hence, they are often used by preference, to minimize future energy expenditure on weeding.

There is some evidence that energy expenditure through the use of tools is, whenever possible, reduced to a minimum. For example, when land is cleared for cultivation, large trees may be left uncut. This saves energy, in one of the most arduous tasks. At the same time it provides shade for part of the farm. Shade is desirable for the farm workers and also for certain crops (such as cocoyams).

If a tree must be felled, the task is often performed with minimum effort by previously 'ring-barking' it, or killing the tree in advance. After felling, the branches may easily be cut (using machetes) for fuelwood. However, cutting up the trunk (or bole) would still

Table 6.4 Examples of traditional crop protection in south-east Nigeria.

Crop type	Protection method												
	Trapping mammals	Picking insects	Inter-cropping maize	Removal of damaged parts	Application of stem-dip	Shaking vine stakes	Application of oil palm wastes	Chasing birds	Application of wood ash	Exposure to salt water	Weeding	Weed preservation	Application of pesticide
cassava	X		X		X		X	X					
yam	X			X			X						
maize	X					X							
melon	X	X		X									
pepper		X						X					
coconut										X			
pumpkin			X									X	
greenleaf									X				
plantain											X		X
tomato		X		X									
okra		X											
cacao		X											
sugar cane													X

Source: Phillips-Howard 1980a.

require great energy expenditure. Frequently, this is avoided by leaving the remainder lying on the ground. It is simply left to decompose but, as it does so, it releases nutrients to the soil. Hence, a saving in energy is combined with a means for supplementing nutrient supply. As Moss (1977) observed, with reference to stumps and roots; this supplement could be important, especially after the nutrients released by burning have been taken up.

Evidently, although the available tools are simple, their use may be highly adaptive and ingenious. The importance of such alternatives is particularly obvious in the case of crop protection.

Alternative methods of crop protection

In terms of energy use, crop protection includes all the means for minimizing losses of potential food energy. These are directed towards the organisms (pests, weeds, fungi and bacteria) which compete with people in their consumption of crops.

There are many methods of crop protection in south-east Nigeria. The two general ones, burning and mixed cropping, are a product of the bush-fallow system. The practice of burning the ground after it is cleared effectively kills vulnerable pests (such as ants and termites) at the beginning of the growing season, whereas mixed cropping, that is, growing a variety of crops together, tends to restrict the spread of pests and diseases (Igbozurike 1977, Taylor 1977).

In addition to these there are many more specific methods of crop protection. Table 6.4 shows that the trapping of mammals (grass-cutters, monkeys and bushrats, for example) and picking of insects (caterpillars, snails and grasshoppers) are common ways of removing pests. At most they involve the construction of simple devices from local materials. These methods actually involve a net input of energy, because the caught animals are eaten.

We have unconfirmed (verbal) evidence that intercropping of maize with cassava and pumpkin helps to reduce pestilence. This may be because the maize attracts pests away from the other crops. The removal of damaged parts, shaking yam vines and chasing birds are highly appropriate means of pest control, some of which are carried out by children. In these examples no tools are necessary. Stem-dips (concoctions of herbs, bark, native gin, fermented palm-wine and camphor), wood ash, saltwater and oil-palm wastes are indigenous pesticides and/or fungicides. They are simply manufactured either from local materials or from materials which are readily available in their useful form. Use of a commercial pesticide (Gamalin 20) was only reported for the protection of plantains and sugar cane.

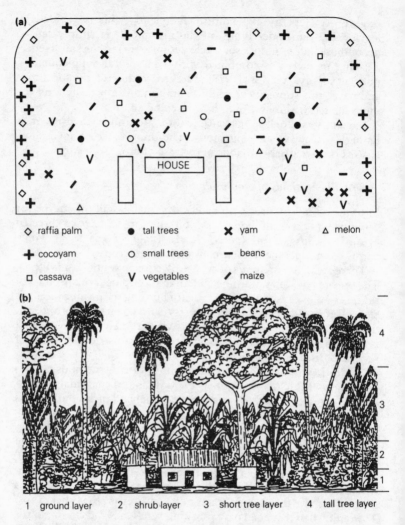

Figure 6.5 A kitchen garden in Obi-Oma Ngwa, Imo State: (a) plan view; (b) side view (from Phillips-Howard 1980a).

Weeding serves not only to reduce competition for solar radiation, nutrients and water, but also to destroy the habitats of insect pests. For example, Barker *et al.* (1977) and Page and Richards (1977) show that the removal of the weed *Eupatorium odoratum* by farmers in southern Nigeria helps to combat grasshopper infestations. Our unconfirmed survey results indicate that the preservation of weeds close to pumpkins may divert pests away from this crop. If this is correct, it is yet another highly adaptive, energy-saving method of pest control.

Energy use in mixed cropping

There is considerable evidence that mixed cropping often results in higher and more dependable yields than sole cropping or mono-cropping (see, for example, Webster & Wilson 1966, Norman 1968, Bradfield 1972). In other words, mixed cropping involves greater conversion of solar energy into food energy.

In south-east Nigeria, the number of crops grown ranges from about 4 to 62 (Okigbo & Greenland 1976). Figure 6.5a is a simplified plan view of a typical kitchen (compound) garden in Obi-Oma Ngwa Local Government Area, Imo State. Such a complex assemblage of crops represents a wide spectrum of micro-ecological conditions, similar to that of the tropical rainforest (Igbozurike 1977). It is managed by the application of simple tools and detailed ecological knowledge to maximize (within the technological limits) the conversion of solar radiation into food. This is achieved by the spatial and temporal organization of crops in relation to the spectrum of ecological conditions. Such organization is manifest in (1) vertical stratification, (2) areal patterns, and (3) crop substitution.

VERTICAL STRATIFICATION

Figure 6.5b shows that the kitchen garden includes tall trees, smaller trees, bushes, vines, cereals and ground-level crops. Hence at least four vertical strata can be differentiated. Such stratification results in an extraordinarily large leaf-area per unit area of ground. This means that solar radiation is absorbed efficiently and the potential for crop photosynthesis is enhanced (see Herrera & Harwood 1973, Igbozurike 1977). Hence vertical stratification helps to maximize the utilization of solar energy in crop production.

AREAL PATTERNS

Patterns of mixed cropping vary throughout tropical Africa and are liable to change (Okigbo & Greenland 1976). In south-east Nigeria they also vary according to the units of organization (that is, compound garden, farm and village).

In the context of energy-use management, the patterns at the village level are most significant. Frequently, the space around the village is organized into three zones (see Fig. 6.6). Zone I, the nearest to the village, includes the compound gardens. Energy is saved because some of the most bulky crops, which require great

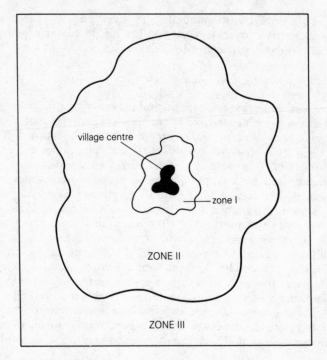

Figure 6.6 Simplified aerial pattern of the villages.

energy expenditure in transportation (such as plantains and yams) are often grown here. An energy saving is also made by locating the crops which are most frequently needed (that is, the soup vegetables) in this zone. In energy terms it makes sense to produce as much food as possible near to the village. This is achieved partly by stimulating production in Zone I through the application of domestic refuse and night soil, and partly by minimizing production losses. Since, as we noted earlier, crop protection depends largely on the presence of people, some of whom are always present around the village, the lowest production losses probably occur in this zone.

Zone II includes most of the temporary farms. Again, they are located as near as possible to the village, but outside Zone I. In between the cultivated farms are abandoned (fallowed) farms at various stages of regrowth. Hence, this zone is a 'mosaic' of intensively managed and little managed areas. Being a zone of intermittent and localized activity, it probably has intermediate production losses. Most fuelwood is collected in this zone, either from the newly cleared or the abandoned farms. The fuelwood and agricultural produce are transported, with minimal effort, when the farmers return to the village.

Finally, Zone III represents the surrounding forest. This is the zone of minimum activity and, if farms exist, the zone of minimum crop protection. Travel between it and the village requires more energy expenditure. In some cases, this is avoided by the construction of huts in which farmers can live temporarily. An additional energy source in the form of bushmeat may be exploited in this zone.

CROP SUBSTITUTION

Crop substitution is a strategy which extends, for as long as possible, the period of food production in a particular area. In energy terms it helps to maximize the conversion of solar energy into food energy by prolonging crop photosynthesis. Crop substitution involves the replacement of one crop by another, so that the second matures after the first has been harvested. This is simply done with the knowledge that crops mature at different rates, and have different requirements which vary with time.

For example, maize is often planted near to yams, even though it may not yield well, because it provides conditions suitable for the growth of the latter. By the time the maize is harvested, three to four months after planting, the yams are well established and their increasing light (energy) and space needs can easily be satisfied. This form of crop substitution not only prolongs production, but also spreads the labour requirement (energy expenditure) throughout the year. Thus it involves efficient management of both inputs and outputs of energy.

Crop rotation is another form of substitution. In this case, crops are replaced on an annual (or sometimes biannual) basis. A simple example is shown in Table 6.5. It is well known that rotations prolong production by maintaining the supply of essential

Table 6.5 An example of a crop rotation in Buguma, south-east Nigeria.

Year	Plot			
	1	2	3	4
1	yams	beans	cassava	maize
2	maize	yams	beans	cassava
3	cassava	maize	yams	beans
4	beans	cassava	maize	yams

Source: Phillips-Howard 1980a.

nutrients. This is done by alternating crops with different nutrient requirements (and root depths) and by the inclusion of nitrogen-fixing crops such as beans. Rotations are also energetically efficient because they increase and prolong energy conversion into food. They also lessen the frequency of land clearance and so reduce energy expenditure in the most arduous task.

Conclusions

In south-east Nigeria the rural communities depend mainly on non-commercial energy sources, that is, naturally converted solar radiation. Since many of their tasks involve human labour, food energy is the most vital energy input. The significance of this has not been fully recognized in studies of rural energy systems. The supply of food energy is achieved by the knowledgeable organization of many crops in relation to a great diversity of managed ecological conditions. It involves the use of a few simple tools with minimal expenditure of energy. This system of mixed cropping is both energy saving and finely adapted to environmental conditions.

However, in the process of Nigeria's oil-based urban–industrial development with its spatial concomitant of rural–urban migration (Mabogunje 1970) and its neglect of the rural areas (Olatunbosun 1975), a visible decline in agriculture has occurred (Floyd 1980). Consequently, although traditional agriculture meets local subsistence needs, it is not sufficiently productive to feed the rapidly growing urban population. Therefore it is necessary to increase the agricultural output.

Since attempts to introduce energy-intensive industrial agriculture in Nigeria (Floyd 1980) and the forest zone of West Africa (Ahn 1974) have generally failed, an alternative strategy must be sought. The results of this chapter support the suggestion by Agboola (1977) that an attempt should be made to upgrade traditional agriculture, rather than to replace it with alien systems. In so doing, full advantage should be taken of the ecological knowledge of traditional farmers and the energy-saving and environmentally sound characteristics of mixed cropping. This may be achieved by implementing an ecodevelopment strategy involving needs-orientation, self-reliance, participation and appropriate technology. It must be said, however, that such a strategy implies further research on energy and ecological aspects of traditional agriculture. Such an approach should lead to increased production and productivity by means which are not only environmentally sound, but also culturally and socially acceptable.

Notes

We would like to thank Professor U. E. Simonis and Dr J. Gaile for helpful comments on an earlier draft of this chapter. This work is a revision of 'A technological alternative for energy use in rural development: the case of southeast Nigeria.' International Politics and Policies, Winter 1984 **5**(1), 16–30, a special issue of the Journal of Public and International Affairs.

1 This survey was conducted from March to May 1980, while Kevin Phillips-Howard was at the University of Port Harcourt, Rivers State, Nigeria. It was mainly intended to gather information on traditional food production. With the aid of students, 56 structured interviews were conducted throughout Rivers State, and 29 were administered in the neighbouring states (Imo, Cross-Rivers, Bendel and Anambra).
2 Ojo (1980) found that, in the city of Ife, kerosene was the preferred cooking fuel. This energy source does not appear to be so important in rural south-east Nigeria.
3 A possible exception is the provision of primary schools which, under the Universal Primary Education (UPE) scheme, are now available in most villages.

References

Agboola, S. A. 1977. *The spatial dimension in Nigerian agricultural development.* Inaugural Lecture no. 22, University of Ife.
Ahn, P. M. 1974. *West African soils.* London: Oxford University Press.
Auswertung dreijähriger Erhebungen in neun biologisch-dynamisch wirtschaftenden Betrieben. 1977. Ministry of Agriculture, Baden-Württemberg (FRG) in co-operation with Forschungsring für Biologisch-dynamische Wirtschaftsweise, Darmstadt (mimeographed).
Ay, P. 1980. Fuelwood and charcoal in the West African forest: field research in western Nigeria. *Proceedings of the first workshop of the UNU Rural Energy Systems project*, 26–38. Ife Nigeria, United Nations University.
Barker, D., J. Oguntoyinbo and P. Richards 1977. *The utility of the Nigerian peasant farmers' knowledge in the management and monitoring of agricultural resources.* General report no. 4. Monitoring and Assessment Research Centre, Chelsea College, University of London.
Biswas, A. K. and M. R. Biswas 1975. Energy and food production. *Agro-Ecosystems* **2,** 195–210.
Bradfield, R. 1972. Maximizing food production through multiple cropping systems centred on rice. In *Rice, science and man*, 143–163. Los Banos, Philippines: International Rice Research Institute.
Floyd, B. N. 1980. *The contemporary agricultural scene in Nigeria: problems and prospects.* Department of Geography, University of Calabar, Nigeria (mimeographed).
Hampicke, U. 1977. *Landwirtschaft und Umwelt.* Doctoral dissertation, Technical University, Berlin (D83/Nr 24).

Hauser, J. A. 1975. Zur Energiebilanz unserer Landwirtschaft. *Neue Zürcher Zeitung*, 12 November.

Herrera, W. T. and R. R. Harwood 1973. *Crop interrelations in intensive cropping systems*. Los Banos, Philippines. International Rice Research Institute.

Igbozurike, U. M. 1977. *Agriculture at the crossroads: a comment on agricultural ecology*. Nigeria: University of Ife Press.

Imevbore, A. M. A. 1973. *Man and environment: the Nigerian situation*. Inaugural Lecture no. 5. Nigeria: University of Ife Press.

Keller, E. R. 1977. Biologischer Landbau – Alternative oder Denkansatz? *Neue Zürcher Zeitung* 26 April.

Leach, G. 1976. *Energy and food production*. Guildford, Surrey: IPC Science and Technology Press.

Lockeretz, W., R. Klepper, B. Commoner, M. Gertler, S. Fast, D. O'Leary and R. Blobaum 1975. *A comparison of the production, economic patterns and energy intensiveness of corn belt farms that do and do not use inorganic fertilizers and pesticides*. CBNS-Ae 4, Washington University, Saint Louis, Missouri.

Mabogunje, A. L. 1970. Systems approach to a theory of rural–urban migration. *Geographical Analysis* **2**, 1–18.

Madon, G., L. Robineau, H. de Lauture and D. Fall 1979. Energy requirements of rural clinics in the tropical zone and renewable energy. *African Environment* **3** (3–4), 97–106.

Makhijani, A. 1976. *Energy policy for the rural Third World*. London: International Institute for Environment and Development.

Makhijani, A. and A. Poole 1975. *Energy and agriculture in the Third World*. Cambridge, Mass.: Ballinger.

Mascarenhas, A. C. 1980. Food production: the total environment and rural development. In *Africa in transition – urban, rural and environmental problems*, J. Bugnicourt and B. Glaeser (eds), 83–92, *Vierteljahresberichte* **79**.

Morgan, W. B. 1978. Development and the fuelwood situation in Nigeria. *Geojournal* **2**, 437–42.

Morgan, W. B. 1980. Rural energy production and supply – with special reference to southwestern Nigeria. *Proceedings of the first workshop of the UNU Rural Energy Systems project*, 10–13. Ife, Nigeria. United Nations University.

Moss, R. P. 1977. Relations between tropical moist forest, edaphic factors and husbandry systems under limiting conditions in tropical West Africa. In *Land use and development*, P. O'Keefe and B. Wisner (eds), 6–17. African Environment Special Reports no. 5. London: International African Institute.

Newcombe, K. 1975. Energy use in the Hong Kong food system. *Agro-Ecosystems* **2**, 253–76.

Norman, D. W. 1968. Why practice intercropping? *Samaru Agricultural Newsletter* (Ahmadu Bello University Zaria, Nigeria) **10**, 107–16.

Ojo, G. J. A. 1978. Energy options for rural planning and development in Nigeria. *Proc. 21st Ann. Gen. Conf., Nigerian Geographers Assoc.* University of Jos, 169–82.

Ojo, G. J. A. 1980. Introduction: search for a rural energy strategy. *Proceedings of the first workshop of the UNU Rural Energy Systems project,* vii–viii. Ife, Nigeria: United Nations University.

Oke, O. L. 1979. Traditional food processing in Nigeria. *African Environment* **3** (3–4), 121–32.

Okigbo, B. N. and D. Greenland 1976. Intercropping systems in tropical Africa. In *Multiple cropping, American Society of Agronomy Special Publication* no. 27, 63–101.

Olatunbosun, D. 1975. *Nigeria's neglected rural majority.* London: NISER/ Oxford University Press.

Page, W. and P. Richards 1977. Agricultural pest control by community action: the case of the variegated grasshopper in southern Nigeria. *African Environment* **2/3,** 127–41.

Phillips-Howard, K. D. 1978. An ecological perspective on energy policy with special reference to rural Nigeria. In *Proceedings of the first national conference on energy policy.* Jos, Nigeria.

Phillips-Howard, K. D. 1980a. A survey of traditional resource management in southeast Nigeria. Unpublished.

Phillips-Howard, K. D. 1980b. Endogenous natural science: a valuable resource for Nigeria's development. *Educational Digest* **2**.

Pimentel, D., L. E. Hurd, A. C. Bellotti, M. J. Forster, I. N. Oka, O. D. Sholes and R. J. Whitman 1973. Food production and the energy crisis. *Science* **182,** 443–9.

Reddy, A. K. N. 1979. *Alternative and traditional energy sources for the rural areas of the Third World.* Paper presented to the Forum on Third World Energy Strategies and the Role of Industrialized Countries, Royal Institution, London.

Ronnenberg, A. 1972. *Untersuchungen zur Wirtschaftlichkeit der biologisch-dynamischen Landwirtschaft.* Darmstadt: Forschungsring für Biologisch-dynamische Wirtschaftsweise.

Sachs, I. 1979. Research for strategies of transitional development from maldevelopment to development. *Ecodevelopment News* **10,** 3–7.

Studer, R. 1978. Energiebilanz der Schweizer Landwirtschaft. *Neue Zürcher Zeitung* 12 October.

Steinhart, J. S. and C. E. Steinhart 1974. Energy use in the U.S. food system. *Science* **184,** 307–16.

Taylor, T. A. 1977. Mixed cropping as an input in the management of crop pests in tropical Africa. *African Environment* **2/3,** 111–26.

Vogtmann, H. 1977. Grundsätzliches zum ökologisch orientierten Landbau. In *Alternative Landwirtschaft.* Vienna, Boku-Arbeitskreis Ökologie der Hochschülerschaft, Universität für Bodenkultur.

Webster, C. C. and P. N. Wilson 1966. *Agriculture in the tropics.* London: Longman.

7 Theory and methods of ecofarming and their realization in Rwanda, East Africa

KURT EGGER and
BRIGITTA MARTENS

The need for new orientation in rural development

International development and aid efforts have been very successful in many cases. Nevertheless, disparities between rich and poor countries and hunger in the Third World are still increasing. Population growth in developing countries appears to neutralize progress. Today's global food production could supply the present world population, but structural preconditions for equitable distribution have not yet been fulfilled, and there is general consensus that food aid cannot be a solution to this problem.

An optimistic, growth-oriented attitude towards technological progress and industrial development was a fundamental part of Green Revolution strategy in the first decade, focusing on technology transfer and extension of innovations to overcome underdevelopment. But, in the meantime, this philosophy has been called into question because exponential growth in industrial production has reached its limits and the negative external effects of growth have begun to accumulate. Moreover, underdevelopment is clearly seen to mean more than lack of technology.

Increasing industrialization, production, consumption, pollution in the so-called developed countries, and population growth accompanied by greater needs left unsatisfied in the developing world means that growth-oriented behaviour will automatically imply *an increase in socio-economic disparity between the developed and the developing countries and greater dependence of developing countries on the industrialized world.* For the developing world, this means two things: (1) the growth of social and economic marginalization and (2) increasing pressure upon all resources. Ever more people and growing needs per capita as results of better education, health care and social welfare, and improved infrastructure are good reasons for raising production rapidly. The number of economically marginal households is increasing, as young families are not able to create new production units rapidly enough. Good, fertile soils are being exhausted. Poor or marginal lands are frequently taken over for cultivation, adding to environmental degradation. This leads to the conclusion that the environmental crisis in the industrialized countries, stemming from overproduction and pollution beyond reasonable levels, is directly related to the cycle of poverty through which the productivity of agricultural areas is destroyed.

Complex and vulnerable systems like human societies require relatively sophisticated approaches to improve their overall situation. Therefore, the newer orientations (Saint & Coward 1977) attempt to harmonize social and ecological conditions and goals. World environmental factors and phenomena related to agriculture are documented in detail in *Global 2000* (1980). The basic interactions between environmental elements and negative impacts as a result of agricultural activity are shown in Figure 7.1. Obviously, in the developing countries, considerable growth in production and economy is required to achieve acceptable standards of living. However, under the pressure of continuing population growth this is not very likely to occur; nor is the implicit expectation that oversaturation in industrialized countries be diminished very realistic. Ecological alternatives in tropical agriculture, for instance, must take into account the limits of productivity of tropical ecosystems, as argued by Weischet (1977, 1984), Ellenberg (1979) and other ecologists, and alternative policies and practices must be strictly guided by these constraints. To achieve increased production, therefore, we can only attempt to approach cautiously the limit of sustainable ecosystem performance.

The theory of unlimited growth is being abandoned more frequently in favour of the theory of ultimate homeostatic balance in the biosphere (Vester 1980), one of the key assumptions underlying the philosophy of ecodevelopment (Sachs 1980), agroforestry (von Maydell 1978, 1982, Combe 1982, Nair 1984), the farming systems approach (Lagemann 1977, GTZ 1982), and many other new alternatives (e.g. Augstburger 1984, Sommers 1983, Werf 1983, Egger & Blezinger 1986). A shift in attitude has obviously taken place; exploitation of nature has now become a relationship based on respect for and maintenance of the natural environment. This change is tantamount to what in biology would be a transition from parasitism to symbiosis.

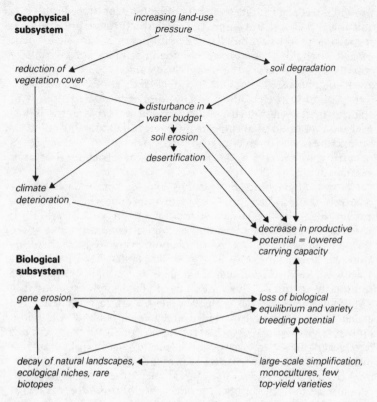

Figure 7.1 Mutual interference of central environmental problems affecting agrarian productivity (from Egger 1983).

Towards an ecological perspective in tropical agriculture

Since the United Nations Conference on the Human Environment (UNCHE) in Stockholm in 1972, the development–environment–agricultural production triangle has been at the centre of many debates on developmental policy. Whereas the concept of the Green Revolution was fixed on agricultural production increase, ecodevelopment aims at balanced, locally adapted development, taking equally into account social, environmental and productive factors. Special emphasis is placed on the necessity of long-term sustainability and the avoidance of irreversible damage to the environment. Within this context, ecofarming, which can be regarded as a highly developed agro-forestry system, is especially concerned with the relationship between agricultural production and the environment.

Careful preservation of ecosystems is a key objective in ecofarming in two respects: first, concerning the long-term economic rationality of maintaining agricultural productivity, and secondly, from an ethical standpoint in terms of responsibility for the preservation of a rich diversity of plant and animal species. The ecofarming approach is based on a systematic understanding of these factors and the recognition that in the complex human–ecological system competing aims must be harmonized. Today, there is general international agreement that the preservation of functioning ecosystems is an aim worth pursuing. Farming techniques, however, rarely conform to this agreement in practice. This, unfortunately, is not only true for many modernization efforts, but also for many previously well-adapted, traditional farming systems which have since become detrimental to the environment under the pressure of growing population, poverty and land scarcity. But what can be done as long as no other methods are known that escape the dilemma of increasing production at the expense of nature, or conserving nature at the expense of production? The long-term consequence of short-term utilization which disrupts the ecological balance ultimately will be a loss of productivity. This conflict can only be resolved if production methods can be found which integrate the diverse flora and fauna of agro-ecosystems as instruments of productivity and sustainability (Tschiersch *et al*. 1984).

The general strategy of ecofarming

Based on the dual aim of conservation and productivity, a general strategy can be derived. First and foremost, the conservation of at least small, scattered remnants of natural landscapes with all their various biotopes and species should be respected. In Rwanda, for example, there are three main natural biotopes: mountain primary forest, valley swamps and savannas. Such areas could prove to be indispensable sources of genetic material; they have already been shown to be good sources of native plant species useful in erosion control and green manuring. So, although it should be possible to utilize these resources under the full constraints of nature conservation, this lies outside the competence of local peasants; it requires at least the political solution of land-use planning.

But preservation of diversity is also one element of conservation strategy for cultivated areas. We can distinguish between two types of diversity: *species* diversity and *biotope* diversity. The latter can

Table 7.1 Hierarchy of methods in ecofarming.

Method	Area found
(1) ECODESIGN★	
scattered trees with bush and herbaceous layer (graded from tree savanna to mountain forest)†	Pare mountains, Tanzania Kikuyu, Kenya
arrangement of contour-parallel erosion-protective lines and interspaced fields; terrace formation†	Bamiléké, Cameroon
(2) ECOPHYSIOLOGY	
(a) *Peaks of endangerment* *Direct programmes*†	
overgrazing ★stable feeding	
erosion damage ★erosion control measures	
decline in soil fertility ★soil improvement	
(b) *Regenerative measures*	
★tree integration	Kikuyu, Indonesia
★green manuring	Usambara, Cameroon
★integration of animal husbandry	Kikuyu
organic fertilizing (dung, compost)	common usage
organic mulching (cut material)	Usambara
live mulching	IITA, Nigeria
(c) *Productive measures*	
mixed cropping	common usage
rotation	local practice
multistorey cropping	Kilimanjaro
rural forest	
(d) *Measures exceeding single farm competence*	
afforestations	
(3) SUPPORTING MEASURES	
mechanical aids	horticulture
mineral fertilizing (I) (raw phosphate, ground lime)	biological farming
mineral fertilizing (II) (NPK)	common usage
plant protection	integrated pest management

Source: Modified from Egger 1982a.
★ The determination of stand structure.
† Indicates elements of indispensable significance.

provide the basis for the former through skilful organization of artificial, mosaic-like patterns with as many ecological niches as possible; these can obtained, for example, by creating ecotones, through small forests, hedges, intensive fallows and mixed cropping, and by tolerating weeds in cultivated areas. Species diversity should now be high, but requires of course careful management and control in order to protect those species which are not directly part of the production scheme. This means that even where intensive agriculture is practised, conservation is more than simply safeguarding the basic factors of production such as soil, nutrients, humus, the water balance and erosion control; it involves living beings as well. We can, therefore, formulate four main goals of ecofarming: (1) ensuring the basic ecophysiological functions, (2) a multiplicity of biotopes, (3) preservation of high species diversity, and (4) the creation of aesthetically pleasing landscapes. Three economic rules are also respected in ecofarming: (1) security is given priority over maximization, (2) subsistence is given priority over commercialization, and (3) farm resource utilization is given priority over external inputs ('low input agriculture' is favoured).

Once a situation in a given project area is sufficiently well understood, it is useful to follow the hierarchy of methods shown in Table 7.1. This is similar to the 'diagnosis and design' method proposed by the International Centre for Research in Agroforestry (ICRAF) in Nairobi. The main ideas are that initially an 'ecodesign' suited to the site must be established; secondly, the ecophysiological setting for employing specific cultivation methods must be identified, based as far as possible on local resources and ensuring a rich bio-subsystem; finally, useful inputs such as tools or chemicals should be selected according to how well they fit into the ecological concept. If this hierarchy is respected absolutely, agroforestry systems will be the most common result or prevalent form of ecofarming, with the possible exceptions of rice cultivation or cultivation of crops in irrigated arid regions. In order, then, to translate this general strategy into recommendations for a specific area, we must combine traditional or autochthonous knowledge with results of ecological analysis and the methods of modern agriculture.

Methods in autochthonous farming systems

Various examples of autochthonous farming systems in densely populated areas show how local people have managed to convert from a pattern of shifting cultivation to one of intensive, permanent agriculture. Local farmers correctly anticipated the potential efficiency of tropical organic farming, and in some cases, sufficient food supply for up to 400 people per square kilometre – even on poor, leached, sandy soils – has been recorded (Lagemann 1977, Ludwig 1979). Three of our own observations made in the East African hill regions (Egger & Glaeser 1975, Egger 1979) will illustrate these systems in detail.

USAMBARA MOUNTAINS

Barely 200 years ago, the Usambara mountains were completely covered by dense forests. They were settled with tribes expelled from the plains by rival tribes shortly before the first Europeans appeared in the country. So the first native invaders only had a very short timespan in which to develop agricultural methods in this new environment. The Europeans, however, were impressed by the local agriculture they found there. From then on, the colonists' farms developed side by side with native methods, which were gradually changing to permanent crops. An attractive landscape was created, rich in forests and with trees integrated into the cultivated areas. This happened, with the arrival of the Green Revolution (during the 19th and early 20th centuries), but this landscape severely altered within two decades: tree stands were reduced, forests were increasingly attacked, polycultures were gradually replaced by modern maize varieties, requiring intensive fertilization. Coffee and tea, originally overgrown by shade trees, began to be grown in open monocultures.

However, some islands of polycultural farms have remained; they still show a dispersed tree layer that provides shade and timber and is undergrown by subsistence crops in mixed stands. A regular fallow of two to four years is introduced when fertility declines; fallow areas are used for cattle grazing. Even the incorporation of wild plants into the growing pattern is systematic and comprises changing tasks according to the growth stages of the crops: as young plants mainly do not cover the ground, co-existing weeds are tolerated to protect the soil from extreme temperatures, from evaporation and, in case of extreme rainfall, from erosion; in addition, at this stage positive competition between wild and cultivated plants accelerates growth. Shortly before the weeds are about to overgrow the crops they are cut on the surface and left in the field as a protective layer of mulch. Their nutrient content is rapidly recycled into the biological system and, enriched with assimilated nitrogen from decomposing bacteria, becomes available to the cultivated plants which are now in their main phase of growth. A second weed generation gradually builds up and covers the ground during the ripening period and after harvest. To maintain the system these weeds are left until full seed ripeness. Thus, at the beginning of the dry season, the fields are covered with a dense layer of dry weeds, keeping the soil humid, smooth and rich in humus – optimum conditions for the next crop in the following rainy season.

We can learn from this that systematic weed tolerance is an ecological method of fertility maintenance, simultaneously conserving numerous species in these plant communities (Bergeret 1977). It is evident that this principle collapses as soon as clean-weeding as an idealized criterion is introduced, as is now happening through the progress-oriented idea of Ujamaa. Nutrient cycles then have to be substituted by fertilization as compensation for hindering the self-performance of the ecological system.

KILIMANJARO

The lower slopes of Kilimanjaro, with abundant rainfall (1500–2000 mm per year) and nutrient-rich soils formed by the weathering of lava, offer excellent agricultural conditions. Apart from a few modern farms which grow coffee, tea, bananas and vegetables separately, most of the area is settled by traditional small farmers. Wachagga families of five to seven people live on small estates of about one hectare, which are covered with forest-like plants: Grevillea trees up to 20 metres high improve the soil with their falling leaves and provide timber and firewood. Smaller fruit trees, mango, avocado and the annona species in particular, are scattered among them. Younger stems are regularly climbed by yams (*Dioscorea*). Bananas form a lower leaf layer four to five metres high; their fruits constitute one of the main subsistence foods, and their leaves and stems are used for cattle fodder. Below them, appropriate to their environmental requirements, coffee shrubs are grown as cash-crops. They yield sufficient money to buy tools and clothes for the whole family. In the shady parts of the lowest stratum, *Colocasia* (*Araceae*) grows well. Its thickened root can achieve up to three kilos of weight and serves as a main source for starch, and the leaves are used as vegetable and fodder. Maize and beans are planted in the areas more exposed to light, with sweet potatoes and local vegetables scattered amongst them. Thus a multi-storeyed culture is built up in perfect imitation of the original natural structure of virgin forest. No weed or pest problems are known in this system; humus conservation is guaranteed by permanent soil cover of organic material, and erosion is prevented. Tillage is essentially restricted to the installation of planting-holes. Cattle are kept in stables, their manure carefully collected and used for selective fertilizing, mainly for bananas. Additional mineral fertilizing is not required. A well-established commercial system supplies these mountain smallholders with grass from the plains, as the essential component of roughage for the cattle is hard to obtain from forest cultures. Thus some of the nutrients washed down to the valleys by drainage are carried back to the slopes.

Not only do the peasants themselves appreciate their cropping system as sufficiently productive, but they feel primarily that it is much more handsome than modern monocultures. Unfortunately this valuation of living space is very often left entirely out of the consideration of foreign planners.

KIKUYUS IN KENYA

The Kikuyu tribe lives on the gentle slopes of the Aberdeen mountains, north of Nairobi. This landscape is densely populated with up to 350 people per square kilometre; it is arranged with high diversity containing numerous trees. Water courses have cut deep parallel side valleys into the main slope. Thus gentle mountain ridges with steep flanks have been formed and these are under intensive garden-like cultivation. Terraces have been established and fortified by fodder grass lines to control erosion. Any organic material is used for mulching the fields. A basic principle in the arrangement of the crops is to follow contour-parallel lines in alternating rows of maize, beans, sweet potatoes, *Colocasia* and so on. Subsequent plantings are in strict within-row rotation. Cattle is kept in open deep stables and supplied with cultivated fodder plants. Fruit trees, such as macadamia nuts, avocado, mango, banana and other trees (*Grevillea*, *Cupressa* and *Albizzia* species), are scattered within the area; they provide a balanced micro-climate and mild shadow. A locally characteristic tree, *Croton megalocarpus*, is of particular importance because of the excellent humus-forming properties of its foliage. This tree is a favoured element in all hedges which fence the separate areas. Hedges are maintained carefully; they give shelter to an abundance of species and produce wood and organic material. The seeds of *Croton* are collected regularly for the establishment of new hedges. The alternation of hedges with loosely scattered trees results in a park-like appearance of the whole landscape.

Modern extension services tend to eradicate non-fruit trees as well as hedges, thus provoking a lingering devastation of the landscape. Native extension-workers react very critically because they feel that the treeless land will turn into steppe with time. This has been proved ecologically correct by the fate of the governmental coffee plantations in that region: all cover trees had been removed to increase the efficiency of the fertilizers for higher yield. New trees have been planted since it became obvious that extreme climatic conditions, droughts as well as occasional frosts, cause damage equal to the expected yield increases, and that these effects can be prevented by planting shade trees. Incidentally, the replanting programme was not initiated by European advisers, but by the insight of native experts.

Summing up these experiences, we find mixed stands, multi-storey cropping with trees, shrubs and field crops, skilful rotation systems, mulching, selective weed tolerance, integrated animal husbandry and organic soil care to be characteristics of those tropical autochthonous systems. As they developed under locally isolated conditions, they largely operated in a closed manner. Thus as well as giving us many ideas, they can still be improved by being treated as open systems in a substantial as well as informative sense: moderate inputs do not need to be entirely renounced and we can compare and combine different systems (Egger 1979, 1980, Metzner 1980).

The role of natural vegetation for the determination of stand structure

Natural vegetation shows us how vegetational density with high continuity in a particular site can be attained. This continuity is usually based on more or less closed nutrient cycles within the organic material of the system. The disadvantage of these systems is that they do not pursue any output interests, whereas it is our task to develop cultivation methods that allow for regular crop take-out. However, the experience of natural systems is that this take-out should be based on a high level of recycling, imbedded into a generally high biomass production.

Especially in the fragile lowland rainforests, examination of the ecophysiology can help to identify methods of sustainable production systems. Tropical rainforest systems work through slow accumulation of nutritive substances acquired mainly from the atmosphere, through dusts, and build themselves up by a recycling system which relies heavily on the presence of root fungi. The soil itself can be nearly nutrient-free, so that over thousands of years nutrients have been accumulated in the organic material. To preserve stability this structure has basically to be maintained. In order to allow the planting of field crops, the tree density may be somewhat lower and suitable tree species have to be selected. Subcultures can either be annual or permanent crops such as coffee, cocoa or bananas. Figure 7.2 shows a range of forest-like formations (Egger 1978, Egger & Metzner 1980), the suitable ecodesign for

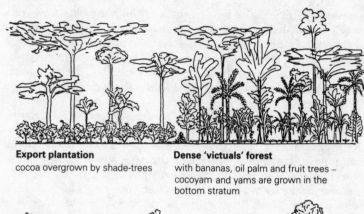

Export plantation
cocoa overgrown by shade-trees

Dense 'victuals' forest
with bananas, oil palm and fruit trees –
cocoyam and yams are grown in the
bottom stratum

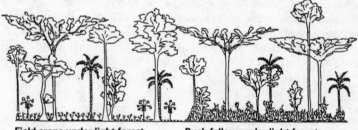

Field crops under light forest   **Bush fallow under light forest**

Figure 7.2 Possible ecodesigns for tropical rainforest conditions (from Egger 1978).

farming in rainforest areas. The role of trees in savanna regions differs greatly from that in rainforests. Water shortage and an extremely delicate water budget restrict possibilities in plant cultivation. It turns out that in the presence of trees productivity can be extraordinarily increased. In contrast to rainforest trees, savanna trees develop deep-growing root systems (up to 25 m) and often reach ground water; due to low rinsing, savanna soils contain comparably more nutrients. In this situation deciduous trees act as nutrient pumps by transporting minerals from lower layers into soil cover. Despite its own transpiration, partly provided by the ground water, the tree of the savanna improves the total water budget and micro-climate by covering and protecting the earth and neighbouring plants from extreme irradiation. Moreover, firewood, as an increasingly rare product, is strongly gaining in economic importance.

Re-establishing scattered tree stands meets with severe difficulties as the short rainy season does not assure sufficient root growth in the vulnerable young stages. Nevertheless, an integrated system of trees, shrubs and field crops has to be aspired to in order to create a favourable micro-climate and soil fertility conditions.

Realization of the ecofarming strategy in Rwanda

The situation of the country

Rwanda is a mountainous African country with altitudes between 1400 and 2600 m, with moderate to steep slopes, formerly covered with forests or, mainly in the east, changing to tree savannas. Five million inhabitants live in 26 000 square kilometres, making Rwanda the most densely populated African country, with an average of 190 people per square kilometre. The annual population increase was 3% between 1972 and 1978.

Geographically there are three main areas to be distinguished: the eastern savanna region with relatively low agricultural potential – altitudes are from 1000 to 1500 m, rainfall of 700 to 900 mm, and average annual temperatures about 21 °C; the central plateau (the project area) with comparably good potential, with altitudes from 1500 to 2000 m, rainfall of 900 to 1200 mm, and average annual temperatures of 14 to 18 °C; and the western hilly area (which separates the watershed of the Zaire and Nile), with altitudes generally above 1700 m to a maximum of 2900 m, rainfall from 1200 to 2900 mm, and average temperature from 14 to 18 °C. Volcanic hilly forest areas more than 3000 m high are beyond agricultural use.

Rainfall is distributed in two rainy periods, the so-called small rainy period from October to December – which can be unreliable and sometimes does not occur – and the great rainy period from January to May, which never fails to occur. In general two harvests are thus assured.

The dominance of cattle raising led to the destruction of most of the forests; huge, highly overgrazed, low-yielding mountain tops are now typical of the landscape. Autochthonous agriculture was characterized by mixed cropping of sorghum, finger millet, sweet potato, manioc and *Colocasia*. During the period of Belgian colonialism, mixed cropping was greatly simplified by the introduction of mono-cropping. The most important permanent crop was and still is the banana. Because of the increasing population density, the arable land is more and more overused. Consequently, soil degradation is in full progress. Ruthenberg (1971) has thoroughly focused on this process of ecological destruction by impoverished smallholders.

Today, 30% of the total area is regarded as suitable for agricultural use, 22% is used as extensive grazing land, 6% is still under natural forest, and the rest is considered not adaptable for farming. The average size of farms declined drastically from 2.65 ha in 1965 to less than 2 ha (1.4 ha in the project area) in 1983. More than 95% of the population depend exclusively on agriculture; they contribute 40% to gross inland production, and 75% of agrarian production is consumed by the peasants themselves. Three-quarters of the export production derives from agriculture; coffee made up 60% of all exports from the country in 1980.

The isolated inland location causes difficulties in access to foreign markets because goods must be routed through the competing countries of Uganda and Tanzania, thus involving risks. There is little urbanization in the country and only a relatively low commercial activity. The land distribution is nearly optimal. Land is owned by the family as long as it is cultivated (usufruct). Communal land, located in the periodically flood valley pastures, is left by the local authorities to the care of farmers for temporary use. People live on isolated farms and have their fields around their homesteads. Hilltops, deserving afforestation, are actually used as marginal pastures and belong to local communities as well. So improvement of the agricultural situation does not rely on land reform but clearly requires a reform in land-use methods.

Project goals

The government of Rwanda is fully aware that sophisticated measures are required to increase production, with methods which deal with the causes of problems rather than symptoms. Soil erosion and degradation, decline in fertility, low production level, low labour efficiency, lack of firewood and so on are interrelated and cannot be solved as separate problems. Central goals put forward by the government are soil conservation, improvement of soil fertility and integration of animal husbandry, in order to overcome the dominant problems of erosion, declining fertility and overgrazing. Figure 7.3 shows an integrated approach which tries to connect these three governmental goals with elements recommended in our ecofarming strategy: tree integration (agro-forestry), intensified cultivated fallow as green manure, and general organic soil management. Interactions between the elements are much more complex than indicated.

Since 1969, the Agropastoral Project of the German Agency for Technical Co-operation has been working in Rwanda; its central station is located at Nyabisindu and the project area extends mainly

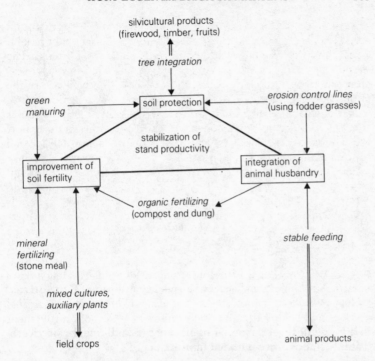

Figure 7.3 Integration of project measures (italicized) and governmental goals (in boxes) in the Nyabisindu, Rwanda, ecofarming project (modified from Egger & Rottach 1983).

over the central plateau. The initial intentions of this project were intensification of agricultural extension work and of veterinary care, increase of agrarian and animal production, and raising the standard of living of smallholders. Ecological difficulties led to complementary objectives which were elaborated in co-operation with the University of Heidelberg in 1975 in the sense outlined above of an ecofarming strategy.

Our aim is sustainable production at a reasonably high level of intensification based on soil fertility and on a diversified regenerating agro-ecosystem with small forests, hedges, trees in the fields and the reduction of grazing areas which are converted to either forests or fields. This ecodesign is illustrated in Figure 7.4.

Elements of the farming system

In the following description of elements of the farming system and their combination, more commonly practised methods are listed

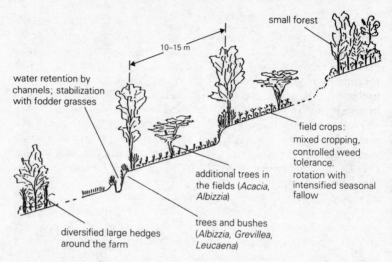

Figure 7.4 Integration of tree lines and fields in the contour-parallel terraces. Small forests and hedges are an integrated part of the farms total biotope mosaic.

briefly, while those of special significance (as indicated by daggers in Table 7.1 above) are outlined in more depth.

INTEGRATION OF TREES

Tree integration on the one hand corresponds to exactly calculable economic criteria; on the other hand, it involves numerous ecological site-relevant aspects which only pay out indirectly, for instance by an increase of field crop yields. Ecological functions of trees mainly ensue from the effects of nutrient transport, fixation of nitrogen, the increase of organic material in the soil, the improvement of soil structure and erosion control.

Trees develop a much deeper root system than annual plants, which enables them to absorb nutrients from deeper soil layers that have been washed down or set free by weathering processes; these nutrients are carried into the biological cycle via wood, leaves and fruits, and so become available to annual field crops.

Integration of legume trees has the additional effect of enriching the soil with nitrogen due to symbiotic rhizobia, associated with the roots of these trees. For many tropical trees, nitrogen is the nutrient they are most deficient in.

Rotting leaves, fruits and roots contribute to the humus content of the soil. In addition, trees with their shade protect the soil surface against overheating and drying. As a result, the promoted activity of soil micro-organisms results in a more rapid transformation (mineralization) of organic material and an increased turnover of nutrients. Increased humus content leads to an improved cation exchange capacity, representing a higher storage capacity of the soil for nutrients. Micro-organism activity also dissolves nutrients fixed in the soil as water insoluble compounds; this is especially common in phosphate, another deficient nutrient.

Tree roots penetrate the soil, break it up and create pores for aeration and watering. Increased water infiltration and improved water retention capacity of the soil benefits plants in case of dry periods during the rainy season. Sometimes the vegetation period can even be prolonged beyond the rainy season. In addition, increased water infiltration reduces soil erosion.

The positive effects of trees mentioned so far can also be attained by scattered planting of trees. In areas endangered by erosion, arrangement of trees in contour-parallel lines is to be preferred, since such an arrangement will reduce superficial water runoff by damming, thus giving time for the water to infiltrate the soil and for transported particles to sedimentate. In the lines, a distance of 3.0–3.5 m between single trees was found to be optimal. Trees standing on their own will not provide a sufficient water barrier. Additional lines of plants along the contour line are necessary, planted close to the tree line and complementing their function. For this purpose, either grass ridges (for example, *Setaria* species) or bush ridges (for example *Leucaena leucocephala* or *Calliandra callothyrsus*) are recommended. In Rwanda, farmers prefer *Setaria* grass as fodder, but three times more biomass is produced by the *Leucaena* line. Also, *Leucaena* reaches a greater age and, as a legume, increases the nitrogen content of the soil. On the other hand, *Setaria* grass lines provide a more effective water barrier than *Leucaena* bush lines.

Runoff water within the fields has to be controlled by means of cultivation methods. Examples are fields planned as ridges (small heaps and ditches parallel to the slope), and complete covering of the soil with densely planted crops and green manuring.

In Rwanda, the traditional method of cultivating sweet potatoes on slopes represents another example: plants stabilize erosion dams along the contour lines. Distance between the dams depends on the inclination of the slope, which also dictates the distance between tree or plant lines as described above.

DIRECT ECONOMIC EFFECT OF TREES

The supply of timber and firewood in Rwanda is deteriorating alarmingly. Wood and charcoal represent the only energy sources. Annual wood consumption is 1 m^3 per capita, adding up to an annual need of 6 million m^3. Even the reafforestation of land not in use for food production and the use of the remaining primary forests would be insufficient to supply the constantly increasing population with firewood. Therefore, fields must also contribute to wood production. As was demonstrated in the Rwanda project, this is possible without any negative effects on the food productivity of the fields. Production of wood and food are not antagonistic goals, but can be combined into a more stable and productive system through the ecological benefits of trees. This was also demonstrated in agro-forestry programmes in other tropical countries (von Maydell 1978). To combine field and forest production, selection of suitable species is of outstanding importance.

Agro-forestial cultivation of trees results in a more rapid growth compared to exclusive forestial cultivation, since trees take advantage of soil cultivation, manuring and the surplus of space and light. In Rwanda, 300–350 trees (*Grevillea robusta*) can be cultivated in erosion control lines on one hectare and will equal one-fifth the number of tress growing on one hectare in a closed forest. Since trees exhibit a threefold productivity in agro-forestry systems compared to forests, three-fifths of the firewood and timber production of a forest can be produced on the agro-forestial site. In other words, 4–5 m^3 of timber and 6–9 m^3 of firewood, representing the annual need of eight people, can be produced on one hectare simply by integrating trees in erosion control lines (Neumann 1984).

In many cases, leaves and fruits of trees can be used as fodder for cattle. Depending on tree species, function and growing conditions, tree pruning can be started as soon as 6–12 months after planting (for example, *Leucaena* species). Branches and leaves of trees can also be used for mulching field crops. The leaf mass harvested annually from one hectare of an adult *Grevillea* stand represents the material necessary for mulching 500–1000 coffee plants (Neumann 1984). Trees also, of course, can produce fruits for direct consumption.

SELECTION OF TREE SPECIES

The selection of tree species is determined by their ecological role and the economic interests of the farmer. Characteristics such as rapid development, quality of timber, suitability of leaves and fruits as fodder or food, as well as low competition for nutrients with field crops, are all relevant. Root pruning (the cutting of superficial roots) does not only limit competition with other plants but also stimulates development of deep-growing roots in *Grevillea robusta*, for example.

In Rwanda, the following tree species were found to be suitable for tree integration into agro-forestry systems: *Grevillea robusta*, *Albizzia* species, *Cassia spectabilis*, *Leucaena leucocephala* (which can be used as a tree or shrub), *Markhamia lutea*, *Acrocarpus fraxinifolia*, *Newtonia buchananii*, *Croton macrostachius* and *Cedrela odorata*.

Suitable for pruning are *Vernonia amygdalina*, *Leucaena leucocephala* and *Cassia spectabilis*. *Sesbania sesban* is a rapidly growing species which can be cut early and replaced by other species. Not only a diversity of species, but also a diverse age structure within the tree stand is desirable for continuous supply. *Miletia dura* is a slowly growing species which provides only light shade and can be used for the production of hard timber.

HEDGES

Like no other element, hedges contribute to the ecological diversification of a farm. The primary function of hedges is to fence fields, paths and buildings. Hedges can be the habitate of a broad variety of animal species. Ideally they should be composed of different plant species, arranged in three parallel sections: on the outer edge, a dense, protective bush line (in Rwanda, *Euphorbia tirucalli* is preferred); in the centre, a line of trees, and at the inner edge, another bush line (for example *Leucaena leucocephala*, *Calliandra callothyrsus* or the thorny *Ceasalpinia decapetala*). The tree line can be formed by the same tree species mentioned above, including *Croton megalocarpus*. To obtain an equal distribution of light and shade, different tree species of different age have to be mixed adequately.

FORESTS

Apart from large-scale reafforestation outside farming areas, the establishment of small forest units on degraded, non-arable soils, as are often found on a farm's border, could contribute to wood production. Effective and economic support of the young plants on these poor stands can be achieved by so-called 'spot-improvement', which means that organic or mineral fertilizers are applied once and exclusively to improve planting holes directly for the first period of growth. As soon as the trees are able to reach deeper layers, growth without further manuring can be expected, resulting in regeneration of the stand. In Rwanda, optimal regeneration and humus build-up was obtained with combinations of pine trees, trees with quickly decomposing leaves (for example, *Leguminosae*) and trees with slowly decomposing leaves (for example, *Grevillea robusta*).

INTENSIVE GREEN MANURING AND ROTATION

In Nyabisindu, previous investigations demonstrated that intensive green manuring would be the most effective method of increasing soil fertility, since even amounts up to 15 tons of cow dung, plus 100 kg of NPK per hectare do not result in yields comparable to those obtained after a well-run, one-year bush fallow. In addition, bush fallow shows more long-term improvements in physical and chemical properties of the soil. A one-year bush fallow was proven to be more effective than a bush fallow of several years or an annual plant fallow. The aspects which explain the beneficial effects of the one-year bush fallow are biomass production, fixation of airborne nitrogen, deep rooting, erosion control and ecological diversity.

Total biomass production of this fallow type is 20–30 tonnes per hectare; even if parts of this material are removed from the field, the soil is supplied with large amounts of organic material. Nutrients from leaves, fruits and roots become mineralized and available for the following field crop. High turnover of biomass results in increased activity of micro-organisms and healthy, growth-promoting soil life. Increased humus content improves soil structure and results in improved physical properties such as better absorption and retention of water. Due to its content of lignified material, the bush fallow provides a slowly decomposing fraction in addition to quickly mineralized leaves.

The following mixture of legumes can be used for intensive green manuring: *Tephrosia vogelii*, *Cajanus cajan* and *Crotalaria* species. Choice and mixture of plants is such that the nitrogen uptake and nitrogen supply of the soil is optimized. The mixture is sown densely and should be composed of a balanced proportion of all three species when fully grown.

In a later stage of development, some of the fallow plants can develop deep roots and thus show the same effect of deep-layer nutrient break-up and groundwater access in a somewhat lower degree than trees.

Dense root systems and slackened soil structure in fallow fields prevent water runoff from the site. As any water from higher areas flows in, it is forced to seep. Transported soil particles are deposited and contribute to improve the site. Even during severe rainstorms, water excess is scarcely seen at the lower edge of a fallow field.

In combination with tree, grass or bush lines, intensive fallows of one or several years' duration are the most effective erosion control at hand (GTZ 1984). Apart from protecting lower-lying areas from runoff, they can very effectively regenerate even strongly degraded soils and return them to agricultural use.

A certain diversity, resulting in an optimized utilization pattern of resources, is given with the above mentioned mixture of chosen fallow plants. A variety of germinating weeds accompanies the young growth stages; according to experience at Nyabisindu, these are predominantly *Compositae*. They produce root exudates resistant to soil-borne diseases and are considered to be beneficial for the following crop. After a few months, weeds are overgrown by the fallow plants, a very effective way of weed suppression especially in the case of troublesome couch-grass.

This intensified green manuring system replaces former fallow practices which are frequently found in Rwanda and other tropical countries and work more slowly in a *laissez-faire* manner. To achieve a high level of productivity, green manuring has to obtain a fixed place in the rotation. This leads to a rotation pattern with green manuring at the end of a cultivation cycle. Natural stand quality, availability of organic fertilizers and crop requirements determine the rhythm of intensive fallow introduction.

In the project area, a 1 : 1 rotation turned out to be optimal; in case of higher dung and fertilizer availability, 1 : 2 rotation with manure and mineral fertilizing between fallows appears to be suitable. The first two seasons after green manuring can reliably exceed original yields twice or three times, or even more. So even the reduction by 50% of cultivated land, as in a 1 : 1 rotation, is easily compensated for by increased total production, in addition to the use of fallow plants as fodder and compost material.

However, the first introduction of a 1 : 1 rotation on a farm is still a problem in two respects: in the first year, green manuring does not yet show its efficacy, and on very meagre sites growth of the fallow plants themselves can be restricted to an unsatisfying degree. This initial phase should be bridged by extensive use of all available within-farm nutrient resources and, if possible, mineral fertilizers. In the long run, a very moderate application of mineral fertilizer would be sufficient to adjust nutrient balance to a high fertility level (Prinz, 1983).

ORGANIC FERTILIZING

Dung and compost serve as the main organic fertilizing material. As their regular amount is rather limited, they are mostly applied to particularly fastidious crops or to regenerate exhausted fields, as for example in the above mentioned spot-improvement. Nutrients from human faeces can be recycled by using the so-called banana toilet: by placing young banana shoots on former toilet pits, they reintroduce nutrients rapidly and safely into the biomass recycling process.

MULCHING

Covering the soil surface with cut organic material helps to prevent soil erosion, improves micro-organism activity, suppresses weed seedlings and so on. Large amounts of mulching material are required to be efficient and the available material is mostly already consumed by mulching the special coffee and banana crops. As a rule, tree integration, hedges and green manuring significantly contribute to increase production of organic material, so the proportion of cultivated land supplied with mulch coverings can be extended.

MINERAL FERTILIZING

The economic situation of most Rwanda farmers does not allow application of mineral fertilizer, and experience from the project area indicates that lasting improvement of agrarian productivity can be achieved by exclusively organic soil management.

However, an initial application of minerals may help to accelerate biomass production and turnover in the initial stage of conversion towards ecological farming methods. Especially on nutrient poor or degraded soils, as is mostly found in the tropics, this input can be of value for obtaining the self-reproductive high biomass recycling level aspired to.

For economic and ecological reasons, the ecofarming concept gives priority to the use of nutrient resources available in the country, such as stone meals, volcanic soil material, or fertile alluvial soils. These contain indissoluble nutrients which are made available to the plant by an active organic soil regime. Even the application of soluble mineral fertilizer becomes more efficient when the cation exchange capacity of the soil has been increased by higher humus content, so that nutrient rinsing is prevented.

INTERCROPPING

In Rwanda and in other countries, it was demonstrated that different forms of mixed cropping are better adapted to tropical conditions than monocropping (Steiner 1982). One special design, combining traditional elements – sweet potatoes on heaps with intersown sorghum – with two modern elements – cultivation in contour lines and introduction of soybean – was very successful, giving a yield increase of 40–45% compared to monocropping, and simultaneously controlling erosion. This combination, practised in demonstration plots, was tested in different zones of Rwanda by the National Agricultural Research Institute of Rubona (Janssens *et al.* 1984). Other successful combinations are maize, beans and manioc; and pigeon pea, sweet potatoes, maize and soybean. Farmers are usually experienced in handling such mixtures; complex intercropping including bananas, fruit trees and vegetables such as *Amaranthus* species and *Gynandropsis gynandra* is often found near the *ruga* (homestead). Fields outside are still dominated by the monocropping of either cassava, sweet potatoes, beans or sorghum. In these cases a return to the intercropping system is now being recommended. Only in the case of special products like coffee, tea, Pyrethrum and densely planted plantains is monocropping indicated.

ANIMAL HUSBANDRY

Another essential element of ecofarming is the integration of cattle or small ruminants by stable feeding, fodder production and careful manure collection. Large areas degraded by overgrazing can then be regenerated by green manuring. Dung is left in the stable for three to four months ('deep litter'), so ammonia becomes completely absorbed and conserved until it is used in the field. Fodder is collected from contour lines, hedges and special plots with fodder grasses mixed with *Desmodium intortum*.

Farms of less than one hectare are not able to produce enough fodder for cows; sheep and goats are then recommended.

MECHANICAL EQUIPMENT

There are many efforts to increase rural labour productivity in the tropics by improved mechanical equipment. Ecological land-use practices comprise measures of additional labour expenditure, as for example fallow care, which might be reduced by appropriate tools and elementary technical aids.

PLANT PROTECTION

Ecofarming involves the principles of integrated pest management; chemical and biological agents should only be applied if economic loss goes beyond an acceptable level. Emphasis is laid on the avoidance of pest outbreaks by cultivation methods concerning the entire biocoenosis (FAO 1967). So far no plant protective agents have had to be applied in the project area due to the successful stabilization of prey/predator balances and related regulation mechanisms.

The methods outlined above constitute the realization of the ecofarming concept and are not meant to be evaluated independently of one another; they are interrelated in a complex and flexible framework (see Fig. 7.5), which unfolds its multiple impacts on site ecology and the production process most efficiently when applied as a whole system. The ecofarming concept would be totally misunderstood if seen as a more or less arbitrary summing-up of ecologically reasonable measures. But, in accordance with natural regulation mechanisms, the scope is to pick up all factors and aspects

(a)

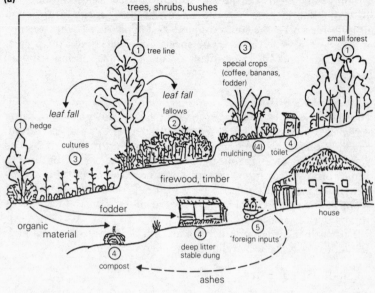

(b)

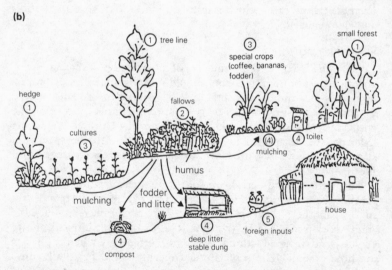

Figure 7.5 Within-farm biomass recycling pathways of two main elements in ecofarming: (a) integration of trees; (b) intensive green manuring by bush fallow.

relevant to the impact of man in his special environment, and to bring them together into a sound, harmonized land-use system appropriate to human and environmental needs.

Note

Fieldwork in Rwanda has been greatly supported by the German Agency for Technical Co-operation (GTZ, Eschborn); actually, after a period of ecofarming identification, all efforts of the project have been concentrated on its extension, and the authors would like to thank all the project members for their co-operation. The assistance of M. Fentzloff in the preparation of the manuscript is greatly appreciated.

References

Augstburger, F. 1984. Erfahrungen mit dem Transfer ökologischer Landbautechniken an Kleinbauern bolivianischer Hochtäler. In *Ökologischer Landbau in den Tropen. Reihe: Alternative Konzepte* 47. P. Rottach (ed.), 183–190. Karlsruhe: C. F. Müller. 2nd edition 1986.

Bergeret, A. 1977. Les systèmes de production écologiquement viables: illustrations dans la domaine de l'agriculture. *Nouvelles de l'écodéveloppement* 3.

Combe, J. 1982. Agroforestry techniques in tropical countries: potential and limitations. *Agroforestry Systems* 1, 13–27.

Committee for Research into Biological Methods of Agriculture 1980. Alternative methods of agriculture. *Agriculture and Environment* 5 1–199.

DSE (Deutsche Stiftung für Internationale Entwicklung) 1982. *Fachseminar standortgerechter Landbau*; Feldafing.

Egger, K. 1978. *Projet de développement agricole du Lekié-Mbam; étude sur l'amélioration des techniques culturales et du contrôle de l'érosion.* Heidelberg: Forschungsstelle für internationale Agrarentwicklung.

Egger, K. 1979. Ökologie als Produktivkraft. In *Agrarreform in der Dritten Welt*, H. Elsenhans (ed.), 217–54. Frankfurt: Campus.

Egger, K. 1980. Ecological evaluation of the situation. In *Factors affecting land use and food production; Reihe Sozialwissenschaftliche Studien zu internationalen Problemen* 55, 132–68, B. Glaeser (ed.), Saarbrücken: Breitenbach.

Egger, K. 1982a. Methoden und Möglichkeiten des Ecofarming in Bergländern Ostafrikas. *Giessener Beiträge zur Entwicklungsforschung* 8, 69–96.

Egger, K. 1982b. Ökologische Alternativen im tropischen Landbau–Notwendigkeit, Konzeption, Realisierung. *Freiburger Geographische Hefte* 18, 119–32.

Egger, K. 1983. *Möglichkeiten für neue Beratungsinhalte auf ökologischer Grundlage. Bericht über Beobachtungen and Untersuchungen im Projekt 'Landwirtschaftliches Entwicklungsvorhaben in der Zentralregion Togo'.* Manuscript.

Egger, K. 1984. L'agriculture écologique, clé du développement agricole? *Développement et Coopération* **3**, 12–15.

Egger, K. and J. Metzner 1980. *Eco-study for Trans III*. Heidelberg: Forschungsstelle für Internationale Agrarentwicklung.

Egger, K. and P. Rottach 1983. Methoden des Ecofarming in Rwanda. *Der Tropenlandwirt* **84**, 168–85.

Ellenberg, H. 1979. Man's influence on tropical mountain ecosystems in South America. *J. Ecol.* **67**, 401–10.

FAO 1967. *Report of the first session of the FAO panel of experts on Integrated Control*. Rome, 18–22 September.

Global 2000 Report to the President 1980. Council on Environmental Quality. Washington, DC: US Government Printing Office.

GTZ (German Agency for Technical Co-operation) 1982. *Projet de ferme familiale pilote de production végétale et animale basée sur un système écologique intégré récupérant l'énergie par recyclage des sous-produits*. Eschborn.

GTZ 1984. *Erosion and erosion control*. Technical note No. 1 of the Agropastoral Project in Nyabisindu.

IFOAM (International Federation of Organic Agriculture Movements). *Ifoam-Bulletin* (quarterly). Kutztown, PA: Organic Gardening and Farming Research Centre.

Janssens, M. J. J., A. Mpabanzi and I. F. Neumann 1984. Les cultures associées au Rwanda. *Bull. Agr. du Rwanda*.

Lagemann, J. 1977. *Traditional African farming systems in eastern Nigeria*. Munich: Weltforum Verlag.

Ludwig, H. D. 1967. Ukara – ein Sonderfall tropischer Bodennutzung im Raum des Viktoria-Sees. In *Afrika-Studie* **22**. Munich.

von Maydell, H. J. 1978. Agroforstwirtschaft – ein Weg zur integrierten Landnutzung in den Tropen und Subtropen. *Entwicklung und ländlicher Raum* **6**, 3–6.

von Maydell, H. J. 1982. Möglichkeiten zur Erhöhung der humanökologischen Tragfähigkeit durch agroforstliche Massnahmen in semiariden Gebieten tropischer und subtropischer Gebirge. *Giessener Beiträge zur Entwicklungsforschung* **8**, 121–30.

Metzner, J. 1980. Fortschritt mit der Vergangenheit: Autochthoner Ansatz zur Stabilisierung eines Agroökosystems in den wechselfeuchten Tropen – das Beispiel Amarasi (Timor). *Die Erde* **111**, 213–29.

Nair, P. K. R. 1984. Soil productivity aspects of agroforestry. *Science and Practice of Agroforestry* (Nairobi) **1**.

Neumann, I. 1984. Vom Nutzen der Bäume in der kleinbäuerlichen Landwirtschaft tropischer Bergländer. In *Ökologischer Landbau in den Tropen. Reihe: Alternative Konzepte* 47, P. Rottach (ed.), 250–62. Karlsruhe: C. F. Müller.

Prinz, D. 1983. Zwei Konzepte zur Steigerung der landwirtschaftlichen Produktivität in den humiden Tropen – die 'Yurimaguas Technologie' und das 'Ecofarming'-Modell. Versuch eines Vergleichs. *Der Tropenlandwirt* **84**, 186–99.

Ruthenberg, H. 1971. *Farming systems in the tropics*. Oxford: Clarendon Press.

Sachs, I. 1980. *Stratégies de l'écodéveloppement*. Paris: Editions Economie et humanisme/Editions ouvrières.

Saint, W. S. and E. W. Coward 1977. Agriculture and behavioural science: emerging orientations. *Science* **197**, 733–7.

Sommers, P. 1983. *Low cost farming in the humid tropics: an illustrated handbook*. Manila: Island Publishing House.

Steiner, K. G. 1982. *Intercropping in tropical smallholder agriculture with special reference to West Africa. Schriftenreihe der GTZ* 137. Eschborn: GTZ.

Tschiersch, J. E., K. Egger, J. Steiger and A. Pfuhl 1984. *Ökologische Problembereiche in Entwicklungsländern und Ansatzpunkte für Projekte der bilateralen Zusammenarbeit*. Forschungsberichte des BMZ **61**. Munich: Weltforum Verlag.

Vester, F. 1980. *Neuland des Denkens*. Stuttgart: Deutsche Verlagsanstalt.

Weischet, W. 1977. *Die ökologische Benachteiligung der Tropen*. Stuttgart: Teubner Verlag.

Weischet, W. 1984. Schwierigkeiten tropischer Bodenkultur. *Spektrum der Wissenschaft* (July), 112–22.

van der Werf, E. J. 1983. Ecological agriculture in Africa – the Agomeda Agricultural Project (Ghana). *Ecoscript* **26**.

8 Yield increase and environmental compatibility through ecofarming

BERNHARD GLAESER

There is clearly a need for agricultural methods appropriate to the tropical environment and which reconcile economic and ecological demands to be identified (see Ch. 7 above). A further goal must be to ascertain the social and economic conditions governing the introduction of ecologically oriented agricultural production and to evaluate how useful they would be for the society of a developing country, in this case, Tanzania. The area of investigation includes the Usambara mountains in the north-eastern part of the country (Glaeser 1984)[1].

The possibilities of increasing income by improving agricultural yield shape the perspective that guided this investigation. The opinions of the peasants in the study area were sought, but in order to assess these and integrate them into policy recommendations, the proposed agro-ecological changes were tested, at first concentrating primarily on erosion control. Because the time available on site was short, the economic evaluation had to be limited essentially to the input factors and to the acceptance of various measures by the peasants and the official extension services (Glaeser 1984, Ch. 6).

Perception of yield increase

To expand on the assessment and to prepare for innovative measures by the competent authorities, it seems important to know how the farmers involved see the potential for increasing agricultural production. This was the intention of the survey, and the catalogue of technical improvements that could be introduced to increase production was thus submitted for evaluation. Specifically, the following matters were examined. First, what were the main farming or agricultural problems in the study area and what solutions were or were to be tried? Following up the results on the study of needs was the second question: how intensive was the increase in production being pursued and for what reason? The question as to how realistic such increases appeared to the interviewees brought up the assessment of methods. Thirdly, the assessment of individual measures should give information about which agricultural techniques the farmers interviewed believed could contribute to an increase in production. The question presented here should provide insight into the level of agricultural expertise. How intensive the accompanying information campaign must be when introducing a package of agricultural methods, for example, depends on the responses to such questions. It would focus in detail on the improvement of labour and capital resources and on technological innovations. In the area of technologies, the purpose was to assess measures for mechanization and irrigation as well as the essential aspects of ecologically sound cultivation. Finally, with this range of methods in mind, the interviewee was to explain as fully as possible which agricultural techniques he thought were the best and to justify his opinion.

The questionnaire was pre-tested by experts, and then discussed with 31 interviewees in ten groups. In the opinion of the experts, the agricultural problems of the investigation area were:

- drought and witchcraft
- infestation with parasites
- soil erosion
- lack of capital
- ineffectual extension service
- inadequate transport

The experts described the problems very much as the farmers saw them. The frequency of the farmers' responses indicated three main problem areas impeding the increase of agricultural production: drought, lack of capital, a decrease in soil fertility, and erosion. In other words, the problems were climatic, socio-economic, and, when environmental influences were considered, agro-economic (see Table 8.1).

Answers to questions about attempts to implement concrete remedial measures revealed a certain helplessness in dealing with the problems. No specific agricultural suggestions were made. The consultations with the rainmaker that undoubtedly occurred in several cases were not mentioned. Only emigration or the search for paid work were reported in a rather fatalistic way as solutions.

Table 8.1 Perceived agricultural problems (in percent of those interviewed).

drought	84
lack of capital	74
loss of soil fertility: erosion	61
infestation with parasites	35
scarcity of land	23
transport	19
seeds	13

In only four of the ten discussion groups was aid from outside sources actually received, specifically, in three villages co-operating with agricultural extension services. Family or friends were reported as supporters in only one group. This may explain why four discussion groups could not imagine that outside assistance was possible. The other groups cited the following organizations and measures:

- the government: credit
- Lushoto Integrated Rural Development Project (LIDEP): transportation
- incorporation into a co-operative: access to state aid
- rainmaker: irrigation

The desire for an increase in production, which was added as a control question, was shared by all those interviewed. The reason was based as much on food as on income. In all, an increase in production was supposed to provide a better life, or at least keep pace with the increase in living costs. The *possibility* of increasing yield, however, was seen as realistic by only two-thirds of those interviewed. Asked for a reason, the optimists suggested the use of certain production factors or the implementation of other measures, such as digging irrigation channels, using manure, fertilizer, pesticides, and better seed; and being more careful in general, such as by regularly pruning the plants. The pessimists said the scarcity of land and the lack of rain and money could not be overcome.

A breakdown of questions into detailed measures showed that irrigation schemes and all types of fertilizers in addition to chemical pest control were considered to be promising. So was most of the package of methods included in the ecological farming approach. Traditional measures such as crop rotation, temporary fallowing, and mulching were acknowledged. The peasants were equally

Table 8.2 Possibilities for increasing agricultural production.

Factor of production	Possibility for increase	Prospects for success[a]
labour input	increase in one's own work	55
	use of hired workers	26
material input	improved seed	90
	manure	100
	fertilizer	87
	pesticides	100
mechanization	plough	26
	draught oxen	16
	tractor	6
irrigation	water bucket	71
	irrigation canals	90
ecological farming methods	shade trees	68
	horizontal anti-erosion strips of grass	87
	mixed cultivation, intercropping	58
	crop rotation	100
	mulching	100
	reduced weeding	10
	leaving land fallow	100

[a] Percentage of the interviewed peasants who judged the measures likely to succeed.

receptive to the suggestion of planting strips of deep-rooting Guatemala grass at regular intervals to prevent erosion. Interestingly enough, some of the interviewees disapproved of intercropping and almost all of them repudiated their practice of tolerating weeds – both measures being traditional. It is still an open question if this response was intended to prove modernity in order to please the interviewer, whether that person be from the Ministry of Agriculture or from far-off Europe. Every type of mechanization was felt to be just as inadequate since the plots of land to be farmed were too steep and too small. Expending more labour, whether one's own or someone else's, was felt to be more or less futile under the circumstances (see Table 8.2).

In a final assessment the groups discussed the 'best' agricultural methods. Articulating an integrated perspective, one group maintained that it was important to care for and work the land well. That included mulching and the process of composting dry material. In addition, they argued that fertilizers and sprays must be used regularly. High-yield crop varieties were considered to be essential. Mulching and regular applications of fertilizer with equal amounts of manure were the most favoured responses, followed by the quality of the seeds used, the application of pesticides, and the loosening of the soil. These suggestions were based on personal experience and on the recommendations of the agricultural extension service.

To summarize, the attitude towards the measures stemmed partly from the farmers' personal awareness of problems and partly from the agricultural methods they have practised. The judgements they made were well founded, and were consistent with the environmental and socio-economic conditions of the study area. The implementation of measures or programmes that deviate from or even contradict the expertise of the peasant would not only be resisted but would, if past experience is an indication, proceed on inherently false or unrealistic assumptions.

Approaches to improving production

A variety of measures for improving production and for increasing productivity were generally accepted as necessary and needed no further discussion. These measures would or should include primarily irrigation and the use of higher-yield, but resistant, crop varieties. It was agreed that pest control and the use of fertilizers must be improved. To a certain extent, these steps are seen as the basic conditions, the *ceteris paribus* variables.

However, an examination of the few less obvious suggestions that this study indicated might nevertheless be purposeful and worthwhile. Such measures primarily concern crop innovations, which help to enhance the productivity per unit of capital. Still more important are the questions about mechanization, since they directly affect the increase of labour productivity. Thirdly, there are at the centre of these considerations the methods that increase soil productivity: ecological techniques.

Crop innovations

Because the population of Lushoto had a rigid scale of preference for its essential subsistence foods and because the cash and export crops of tea, vegetables, and coffee had a secure place, ensuring economic stability in this context must depend only on a few additional crops. From the wide range of Lushoto's less familiar yet more cultivable and worthwhile crops, three have already completed a successful trial period – wine, macadamia and cardamom.

Mechanization

Ecological farming procedures that are appropriate for tropical conditions are one possibility for increasing agricultural production without eventually destroying its ecological foundation. This does not preclude the possibility that more efficient use of human labour might also increase production. These considerations bring up several questions. Which forms of mechanization exist? Which forms can be introduced spontaneously? Which of these forms are consistent with the region's established production patterns and do not threaten the long-term preservation of soil fertility? To answer these questions, it may be expedient to offer a brief overview of the tools and implements in Tanzania itself, as they must be readily available, relatively inexpensive, and must not consume foreign currency. For the present purposes of improving agricultural output, it seems appropriate to categorize the country's existing forms of mechanization according to their function in farming:[2]

- preparing the land
- working the fields
- weeding
- harvesting
- processing the harvest
- transport

PREPARING THE LAND

The hand hoe (*jembe*) and the machete-like *panga* are still the only instruments used on peasant farms. One could improve both by adapting them to the type of work or by using animal power. The first suggestion could involve the use of more durable material. The blade of the hoe, for instance, could be fashioned from more highly tempered steel. Constructive improvements would have to consider various handle lengths, blade widths, grips, and angles of incidence. Single or multiple-pronged ploughs that can be pulled by oxen have already been developed by the Tanzania Agricultural Machinery Testing Unit (TAMTU). The three-pronged plough cost T. Shs. (Tanzanian Shillings) 915/– in 1974. An animal-drawn disc harrow, also developed by TAMTU, cost T. Shs. 980/–. It is doubtful, however, whether teams of oxen can be used effectively in the fields situated on the steeper slopes.

WORKING THE FIELDS

Traditionally, the fields are worked in a very labour-intensive way. A hole is dug, and the seed is placed into it by hand. Hand and animal-drawn implements are widely unknown, although they would save both time and seed. Favourable sowing schedules could be observed more regularly, too.

The same argument against the general use of ox-drawn implements can be used against TAMTU's hand planter developed for the three-pronged plough for T. Shs. 200/– (1974). Hand implements like the one developed by the Lushoto Integrated Development Project (LIDEP) and sold for T. Shs. 45/– in 1976 are certainly easier to be used. The tip of this tool is pushed into the ground whereupon a mechanism releases the seed, which then falls into the soil. This eliminates tedious bending. Labour efficiency soars to between one-half and one hectare of land sown per day. Similar implements, some of which are even simpler (without spring mechanisms), were developed both by TAMTU's Department for Village Technology and the Department of Rural Engineering at the University of Dar es Salaam for sowing corn and beans. The introduction of such tools is worthwhile when there is a lack of labour or when seed prices are high. There are no topographical difficulties. An additional advantage is that subsequent weeding can be done much more quickly if the seeds are carefully sown in rows, which can be assured by using a guide string.

WEEDING

As with preparing the land, one-hand hoes are used for weeding. This chore takes more time than any other. TAMTU's price for a cultivator (a weed plough whose width is adjustable) was T. Shs. 280/– in 1974. This implement can be pulled by oxen or donkeys. Because the LIDEP's donkey programme in the Usambaras was not successful, however, the use of this tool can not be realistically considered.

HARVESTING

Non-motorized implements for mechanizing harvest work are not available. But the labour is not a great limiting factor in this functional domain anyway, because harvesting can be distributed over a long period of time unless the next sowing deadline is pressing.

PROCESSING THE HARVEST

The shelling and milling of corn is the most important activity in processing the harvest. Pounding with mortar and pestle is still seen in the villages. The advantage – the preservation of vitamins and other nutrients – is, however, kept with simple hand milling. A hand-powered maize sheller suitable for household use was sold by TAMTU and LIDEP for T. Shs. 80/– in 1974. There were numerous motor-driven corn mills run as small businesses or as a communal village activity.

Traditionally, most travel like the necessary trip to the various markets is done by foot with the goods carried on the head. The widely advertised ox-drawn or double mule-drawn wagon is certainly less suited to the mountains. Experience on Mount Kilimanjaro has shown, however, that the donkey can be used as a beast of burden. Wheelbarrows and handcarts have also been utilized and have established themselves in certain areas.

EVALUATION

There have been promising beginnings for efficient, appropriate techniques in the agricultural sector, but in general there is far too little technical, financial and training support. There are fairly few implements that appreciably increase the productivity of rural labour without involving excessive capital expenditure. There is a technological need that should be satisfied through innovation within the domestic workshops that already exist. The use of appropriate mechanization in agriculture could complement the ecological improvements and make them more effective.

Ecological techniques

Ecological techniques are central to the ideas and suggestions for increased production combined with resource conservation and environmental protection. They involve a synthesis of traditional African agriculture and European ecological methods (Egger & Glaeser 1975) that have been modified for tropical conditions (see Ch. 7 above). This section presents the initial results of introducing such a pattern of methods in the Shashui area of the Usambara mountains in Tanzania.[3]

EXPERIMENTAL FIELDS

Fields in the village of Msikitini, located near the market town of Soni, were chosen for experimentation planned according to the principles of ecological farming. This was because Msikitini was readily accessible from the local road network and, being surrounded by hills, comprised a relatively closed ecosystem. The experiment was carried out with the co-operation of the native farmers. First, strips of Guatemala grass were planted in contour rows to control erosion. As a further measure, bananas, cassava, corn and beans were intercropped in contour rows to facilitate the

Tabe 8.3   Input structure and costs for an ideal model field.

Inputs	Itemization	Costs per acre (in T. Shs.)	Potential costs
Guatemala grass	600 sprouts (no charge)	–	–
	transport	80/–	20/–
	planting (20 work days at 16/–)	320/–	–
	total	400/–	20/–
bananas	80 trees at –/40 each	32/–	32/–
	transport	40/–	10/–
	planting (15 work days at 16/–)	240/–	–
	total	312/–	42/–
cassava	4000 cuttings = 1000 stalks at –/10 each	100/–	100/–
	transport	900/–	225/–
	planting (8 work days at 16/–)	128/–	–
	total	1,128/–	325/–
grevillea, albizia, avocado	25 saplings at 1/– each	25/–	25/–
	transport	9/–	2/–
	planting (2.5 work days at 16/–)	40/–	–
	total	74/–	27/–
costs for one-acre model field		1,914/–	414/–
costs for one-hectare model field[a]		4,785/–	1,035/–

Note: The calculations pertain to an easily accessible location in Shashui.
[a] 1 acre = approximately 4000 m^2.

work, to make for more even irrigation, and, finally, to ensure against erosion. Guatemala grass is suitable for cattle fodder and so provides the basis for a closed system of mixed agriculture. To support all these measures, a small tree and seed nursery was established on the floor of the valley. Only materials accessible to the native peasants – like the support stakes and shady banana leaves – were used. More Guatemala grass was to be grown along with tree crops such as avocados.

On the basis of the calculations for each of the three experimental fields in Msikitini, the production function of a plot one acre or one hectare large can be inferred (see Table 8.3).

COST PROJECTION FOR THE MEASURES

In principle, it is possible to extrapolate the costs for each village or for the Shashui study area as a whole on the basis of the size given for the plots of cultivated land. The total costs of the required measures can thus be calculated.

The entire area of cultivated land was about 500 hectares. This estimate is arrived at realistically by halving the amount of all cultivated fields reported by the contact persons. On this basis and in keeping with the type of ecologically sound measures implemented and subsequently evaluated in economic terms (which included provision and planting of Guatemala grass, bananas, cassava and various trees), a budget of T. Shs. 2.4 million would be required.

Such a sum may seem utopian considering that entire mountain regions should profit from measures such as these. Even so, it is certainly possible to reduce costs if specific matters are taken into consideration. All pilot projects, for example, include costs of initiating innovations, expenses which can later be eliminated or reduced. Transport of plants can be reduced to one-quarter of the costs reported here by using trucks instead of Land Rovers. Above all, domestic administrative measures should be planned that would appeal to the spirit of self-help by making the materials available only to those farmers or communes that undertook to do the planting themselves. This could be carried out under the supervision of the agricultural advisers, who are responsible for the region anyway. The computation would then be as shown in the final column of Table 8.3, with plant costs remaining the same, transport costs 25% of the previous ones, and labour costs eliminated. This reduces the required external financing to T. Shs. 414/– per acre of T. Shs. 1,035 per hectare, which, with 500 hectares of arable land, would require somewhat more than half a million shillings (T. Shs. 517,000).

Even these costs would be valid only for a crash programme. Naturally, it would make more sense to promote regional independence along with the measures for cultivation by encouraging the expansion of each region's stock of seed. This should be done on a co-operative basis since few peasants have the land necessary for it. How much is saved depends on the size and number of such seed nurseries. The first attempt to increase the seed stock in Shashui is mentioned above. Economic appraisal is contained in Table 8.4.

SCHOOL PROGRAMME

The ecological measures, erosion control, and production improvements described above were disseminated first by the primary schools in the immediate area. The school teachers and the Ministry of Education (*Elimu*) in Dar es Salaam showed interest because an ecological school garden programme complies with the ministry's call for a certain degree of autonomy in the schools. Within the framework of agricultural instruction, pupils and teachers should be

Table 8.4 Input structure and costs for an ideal model Guatemala grass seed nursery

Input	Itemization	Costs per acre (in T. Shs.)
material	6000 sprouts per acre (15000 sprouts per hectare)	–
labour	6000 sprouts cut and carried to the street (2 workers × 1 work day for 500 sprouts) 24 work days at 10/- per day	240/-
transport	6000 sprouts in Land Rover from Lukozi to Soni for Shashui (2 × 40 km at 2/30 per km; 500 sprouts per trip) 12 trips at 184/-	2208/-
labour	6000 sprouts: soil turned up; hole dug; sprouts set in holes; soil packed around roots (2 workers × 1 work day for 500 sprouts) 24 work days at 10/- per day Keeping field clear, weeding	240/- –
costs for establishing a model field of Guatemala grass	one acre one hectare	2688/- 6641/-

in a position to produce the food to meet their needs and possibly even to achieve a surplus to finance the teaching materials.

The five primary schools participating in the ecology programme in 1976 were located in Soni, Kisiwani, Baghai, Shashui and Bumbuli. The school garden programme embraced the steps outlined below (Kimela 1976):

- planting tree crops and banana saplings relatively far apart
- intercropping with various subsistence crops like corn, beans, and high-return vegetables like potatoes and tomatoes
- mulching
- spreading composted manure
- terracing

Because all five schools were located on slopes, each began by digging and building up the one-metre wide terraces across the incline. The importance this had for irrigation and erosion control was also stressed in the instructional programme. The runoff rainwater was channelled from the end of one terrace into the next just as in a canal system. The effectiveness of erosion control can be enhanced by planting deep-rooting Guatemala grass at the edges and on the slopes. The programme also included the use of the grass for mulching material and cattle fodder. The cultivation programme was finally initiated on this basis and run on a plot of land one-half to four hectares in size. Corn, beans, white cabbage, tomatoes, sweet potatoes and onions were planted as needed. The limited use of pesticides was taught especially in connection with vegetables.

The example of the Shashui primary school is presented below and evaluated on a quantitative basis as far as the period covered by the investigation allows. The village made available about four hectares of land which did include a fertile valley floor but was mostly land of little value on the slopes. The incline was terraced in the manner described above. Corn was planted at the edges and surrounded by rings of mulch in which water collected. Tomatoes and carrots were the main vegetables. Table 8.5 presents the farm budget for one year as given by the teacher responsible.

The manure provided by the villagers is not included, nor is the expenditure of labour even roughly calculable in this rudimentary computation. What can be said is that the work needed for terracing cannot be performed by a peasant family, but it could be accomplished in co-operatives or as a communal activity. Above all, it was important for the people to perceive not only that the ecological programme can be carried out, but that it is worth emulating, too. Farmers in Mazukizi, Magila and Msikitini who were following the curriculum of their children were already applying the principles of the school programme to their own fields during the observation phase.

Evaluation of the measures for improving production and increasing yields

The goal is the permanent improvement in the peasant's income, earned in the form of cash and/or means of subsistence. This goal requires the achievement of a consistently high yield over a prolonged period of time. This implies that the fertility of the soil should not be threatened by overcropping and that measures to protect and preserve the environment should be implemented. By so doing, the criterion of harmonizing economy and ecology can be met. A difference between economy and ecology is that the former is short term, the latter long term in perspective. Within the

Table 8.5 Estimated farm budget for the primary school in Shashui (in T. Shs.).

	Corn	Vegetables
gross returns	900/-	1000/-
seeds	83/-	660/-
fertilizer	140/-	
gross margin	1017/-	

long-term scope promoted by this study, the most important role is played by microeconomic factors, specifically those having to do with production theory. The optimal combination of the production factors to be introduced must be found. In order to minimize costs, identified constraints in the factors of production must be relieved.

Since land is clearly scarce and will remain so because of the high rate of population growth, the productivity of the land must be increased. In a region that is overpopulated and that has few job opportunities, the solution would seem to lie in using more labour, either by extending work time or by recruiting additional workers. But since it is evident that labour cannot be increased ad infinitum, the challenge is to increase labour productivity along with land productivity. Both are traditionally achieved through increasing capital input in the form of mechanization. But this is precisely where there is a particularly severe constraint among peasants in developing regions. The dilemma unfortunately so typical of underdevelopment emerges: almost all factors are scarce, but the means to eliminate this scarcity do not exist.

That is why technological measures to increase yield and improve production are being sought, measures that will increase soil and labour productivity with the lowest possible increase in capital expenditure. In addition to the various crop innovations that increase cash income, the main ways to achieve these goals are the introduction of limited mechanization appropriate to the ecological and socio-economic conditions of the respective area and use of a variety of ecological techniques. The combination of these two approaches creates a different kind of production pattern whose main characteristic is that it uses nature and physical environment as production factors and uses ecology as an economic and technical means of production – in both cases more intensely than in traditional farming and chemo-technical agriculture.

Because the experience with ecological measures has thus far been brief, there can and should be only a tentative economic evaluation. Still, erosion control by planting Guatemala grass is a longer-term measure whose agricultural advantages will already be observable in the medium term when it prevents the top soil from being eroded and thus preserves soil fertility. Furthermore, growing Guatemala grass provides an additional reservoir of cattle fodder that can be used to full advantage with stall feeding. The proportion of grazing land that is saved in this way would more than compensate for the amount of land lost in the erosion control strips.

Planting crops like bananas or corn in contour rows does increase labour because planting the cross rows demands more work than simply digging the holes for the plants. On the other hand, planting the crops in rows has various ecological and economic advantages, such as greater density of plants and thereby a greater yield; internal rotation (fallow rows) without losing an entire field (minimization of risk); controlled weeding and mulching, which makes some of the hoeing unnecessary and therefore saves labour; and easier cultivation, possibly supported by small-scale mechanized farming methods.

Intercropping has successfully been applied in traditional farming in that crops with complementary nutrient requirements are planted together in the same field. A quantitative comparison of yield with monocultures is difficult, but certainly the inputs and costs of fertilizers and pesticides decrease in such an ecologically sound system.

In addition to an improvement in yield, in cash return, and in long-term preservation of soil fertility, the optimal combination of production factors is also determined by the chances a crop has of making a profit and the willingness of the farmer to assume risk. The crops grown primarily for home consumption (subsistence farming) tend to reduce farming risk and with it, of course, the profit opportunities. Vice versa, profit opportunities increase with the cultivation of cash-crops, especially those bound for export. At the same time, however, there is a rise not only in risks due to the fluctuation of prices on the world market but usually also to the capital risk linked to the greater volume of inputs.

The decision that regulates the proportion of subsistence production to market production is made according to the needs of the peasant family. Generally, subsistence farming has the highest priority, but the willingness to accept risk increases as personal needs are met.

Notes

1   This contribution was taken from Glaeser 1984, Ch. VI (pp. 161–93) which was abridged and improved.
2   The author draws here on illustrative material obtained during visits to artisans' workshops run by the Tanzania Agricultural Machinery Testing Unit (TAMTU) in Arusha and the Lushoto Integrated Development Project (LIDEP) in Soni. Supplementary information is found in the reports by Macpherson (1975), pp. 116–85 and Tschiersch (1975), pp. 94–125.
3   Professor K. Egger (Heidelberg) and Dr H. W. Schönmeier (Saarbrücken) also participated in this research project, which was financed by the Kübel Foundation (Bensheim) within LIDEP. The Community Development Trust Fund (CDTF) supported the project as the Tanzanian partner organization.

References

Egger, K. and B. Glaeser 1975. *Politische Ökologie der Usambara-Berge in Tanzania*. Kübel Foundation Series. Bensheim: Kübel Foundation.

Glaeser, B. 1984. *Ecodevelopment in Tanzania. An empirical contribution on needs, selfsufficiency, and environmentally-sound agriculture on peasant farms.* Berlin/New York/Amsterdam: Mouton.

Kimela, K. G. 1976. *A survey on ecologically based advice for farming in five primary schools in Lushoto District – preliminary report.* Paper prepared for the Lushoto Integrated Development Project (LIDEP), Soni (mimeographed).

Macpherson, G. A. 1975. *First steps in village mechanisation.* Dar es Salaam: Tanzania Publishing House.

Tschiersch, J. E. 1975. *Angepasste Formen der Mechanisierung bäuerlicher Betriebe in Entwicklungsländern.* Heidelberg: Forschungstelle für internationale Agrarentwicklung.

PART III

Outlook

9 Towards a second Green Revolution?

IGNACY SACHS

The Green Revolution, phase one, was predicated on two contradictions and two misjudgements. Geneticists were asked to breed high-yielding varieties of wheat and rice responsive to massive doses of water and fertilizer. They succeeded in their job, but as the new varieties proved vulnerable they had to add to the production function another costly and often imported component: pesticides. Thus a highly capital-intensive package was offered to solve the problems of peasants and countries starved of capital. At best it could be applied in limited areas, channelling scarce investment resources into capital-intensive agriculture and increasing the dependence on foreign-made tractors, industrial and energy-producing equipment and the supply of fertilizer and pesticides.

When all the exceptional conditions required by the Green Revolution are met – the lucky cultivator has access to water, power, fertilizer, pesticides and 'miracle' seeds – his output is likely to increase substantially. But the bias towards interregional and interpersonal inequality built into the Green Revolution is likely to worsen rather than improve the material situation of the large masses of the rural and urban poor by-passed by the Green Revolution. They will thus continue to starve because their primary problem is not the lack of an adequate supply of food but the incapacity to purchase it. Supply-oriented food policies offer at best a partial solution. The most difficult aspect – how to ensure people's entitlement to adequate food intake – is assumed away. It can be argued that a different kind of genetic research, applicable in less exceptional conditions on a far larger scale, would have done a better job.

Let us turn now to the misjudgements. The Green Revolution was conceived on the assumption that the abnormally low prices for petrol extracted by Western multinationals from Third World oilfields would continue forever. Accordingly it was taken for granted that the cost of fuel for tractors and irrigation pumps, as well as the cost of fertilizer, would not constitute much of a problem. History played havoc with this assumption, the net result being a considerably higher price for food to the consumer (the cost of transportation from the field to the city has also gone up) with a reduced margin of value added accruing to the farmer; the main beneficiaries of the new situation are the suppliers of fuel and fertilizer.

The second misjudgement refers to the severe underestimation of the environmental disruption caused by the Green Revolution package. In a sense, it was bound to happen, because the geneticists were not even asked to consider the environmental aspects in their cost-benefit analysis. Externalities by definition are left out. However, excessive concentrations of fertilizer and pesticides contaminate the streams and the water-table with serious hazards for the population; fish cannot be found in the paddy fields any more so that the farmers' nutritional standards worsen even though their monetary income is on the increase, and so on. More seriously, the genetic pool of plants – the most precious endowment for the future of the human kind – is in danger of being impoverished and partially lost because of the dissemination across the world of a few varieties. The vulnerability of this high-yield agriculture to pests and diseases should not be minimized. Finally, the soils are often ruined by agricultural practices directed towards the maximization of yields in the short run with little or no regard for the preservation of life-support systems. Costly new irrigation schemes are needed year after year just to offset the losses of agricultural land due to salination.

Clearly, the first phase of the Green Revolution, whatever the judgement made on it, is overrunning its potential. To extend it to new areas, even limited, will require enormous outlays of capital. In order to avert complete deadlock, it is high time the Green Revolution was redirected along a new path. We should start by spelling out the relevant criteria for agricultural research oriented towards the solution of the food problem in the Third World.

Food production must be labour absorbing if we want to act simultaneously on the supply and on the demand side. More people must be able to earn their access to food while working to produce it. It is, however, not enough to create more opportunities for employment and self-employment in agriculture and related fields. Labour productivity should at the same time reach a level guaranteeing even to the small farmer an income covering his family's living expenses and allowing for some purchase of industrial goods. Otherwise the development of a national industry would be jeopardized by the lack of an internal market. Moreover, the rural exodus would continue, forcing Third World countries to expend a growing share of their GNP for the infrastructural investment and maintenance costs of the cities.

The two conditions – increased labour absorption and fair labour productivity – are contradictory in the short run, at least in countries which have already exhausted their reserve of agricultural land and are operating at fairly high levels of per hectare productivity. But more agricultural employment with higher agricultural incomes should be within the reach of countries with an active agricultural frontier and those where a considerable gap exists at present between the technically feasible and the actual per hectare output; much will depend in both cases on the right choice of the product mix and of the technologies. Both should lean towards an increased labour intensity. Changing the output mix towards products that require more labour inputs per unit and/or have higher prices might prove an attractive possibility subject to the availability of markets.

In the longer run, to reconcile the two objectives referred to above would require new technological breakthroughs (and new products) calling for higher labour inputs but generating more than proportional increases in the value added per unit of land out of which self-employed farmers and agricultural workers could derive higher incomes. The increase of value added per unit of output can be also achieved by reducing the volume (and/or the unit price) of the inputs.

One of the goals for future research is to devise less resource-absorbing production functions. In this connection, another pair of criteria may be put forward. Food production ought to be less energy-intensive and, at the same time, more energy efficient. There are good reasons to believe that exploring systematically the intersections of the food system and the energy system may help to identify opportunities for rationalizing both simultaneously. Energy recovery from agricultural residues and manure, at one end, and, at the other, energy saving on transportation by increasing local self-reliance in food, point to the broad range of policy options. The United Nations University Food–Energy Nexus sub-programme is investigating the energy profile of the food system taken as a whole from the farm to the consumer's fork (Sachs 1984).

Behind energy conservation and efficiency looms the larger issue of ecological sustainability of the proposed food systems. The goal is to obtain higher yields on a regular and continuous basis, giving up the present predatory resource-use patterns in which part of the production is achieved at the expense of the life support systems. Losses of agricultural land, depletion of aquifers, man-induced desertification and other adverse climatic changes resulting from abusive deforestation are not at present included in the costs of food production. It is only by externalizing these costs (the depletion of the capital of nature) and then assuming them away as if they did not exist that some agricultural operations are presented as a success story, while in reality they should be considered as nothing short of a catastrophe. The most urgent task, as far as agricultural research is concerned, is to internalize the long-term ecological problematic and to give to the concept of sustainability a key role in the evaluation of all new agricultural production functions. The ecodevelopment approach (Sachs 1980, Sachs *et al*. 1981) postulates the search for socially desirable, economically viable and ecologically sustainable solutions.

This is certainly easier said than done. But a number of promising research paths can nevertheless be indicated. Regenerative agriculture, as defined by the Rodale Institute in the USA, is the goal. A much closer look at the resource potential of each specific ecosystem is called for. Food can be produced in a variety of ways. We have lost the ability of past generations to derive a living from the specific flora and fauna of the diverse ecosystems. Rather, an arrogant vision of nature dictates the costly and often wasteful transformation of the environment to make it fit for exotic technologies, as if it were absolutely necessary to make consumption patterns uniform all over the world. Wheat, bread and noodles, without forgetting coca-cola, stand out as symbols of modernity.

This plea for ethno-ecology and historical anthropology of food should not be understood as an invitation to screen the past in order to find ready-made solutions for the future. At best, we can hope to identify some promising starting points for research that should be pursued with all the means of modern science. Genetics and biotechnology should certainly be put to work, not only to produce out of biomass a greater variety of goods and energy but to transform it into an ever wider range of industrial products. The prospect of a new industrial civilization of the tropics based on the uses of biomass should be pursued.

An area of particular concern, still waiting for its neolithic revolution, are the aquatic resources. With few exceptions, men still act as hunters rather than breeders of fish. Fish hunters have been equipped with ocean-going vessels, floating factories and modern weapons for fish destruction. As far as the destruction of marine species is concerned, their war proved quite successful. As for the economic results, they are nothing short of a disaster. Ocean fishing is one of the most energy-intensive operations devised by men. Meanwhile aquaculture continues in its infant stage. A 'Blue' neolithic revolution deserves maximum priority in research and experimentation, both in seas and inland waters. Of all the economic frontiers yet unexplored this offers perhaps the most exciting prospects.

In a sense the first Green Revolution not only transformed the agriculture into a market for industrial inputs, but also applied to food production the industrial philosophy: specialized monoculture became the main thrust of agricultural modernization, the assumption being that it would bring more efficiency. Yet this view can be challenged. The traditional peasants were in fact wiser in that they considered their farms as food-production systems with manifold complementary activities, closed production loops and the transformation of practically all the farm wastes into wealth. Peasants knew out of experience that time-consuming conservation of life-support systems – soils, water, forests – is a necessary condition for their survival and that the fate of their children and grandchildren would depend on their planting trees that will take 50 or 100 years to grow. The traditional peasant farm is based on an empirical knowledge of ecology; it imitates the ecosystem. A major advance is possible today by applying the systems approach to the design of integrated food-energy schemes. Such schemes should of course be ecosystem-specific, so as to make the best possible use of the resource potential of each natural environment, and culture-specific, so as to gain from the knowledge accumulated by peasants along the centuries.

Once more, the issue is not to dig from the past a model, but to combine the peasant rationality with all the possible inputs of modern science. Agricultural systems can range from an improved version of traditional farming to extremely sophisticated, computer-managed agro-industrial complexes, where agriculture and intensive small-scale fish breeding come quite near to this ideal, based as they are on the substitution of material inputs by a combination of knowledge and labour.

Finally a special mention should be made of the prospects held by research scientists for direct fixation of nitrogen by plants. If successful, it would eliminate the need for nitrogen fertilizer – an instance of knowledge-intensive agriculture at its best, in which technical progress is almost disembodied – something all development economists are looking for.

A critical view taken at the first Green Revolution should not lead to a negation of the tremendous potential held by agricultural research, but to a radical reappraisal of its objectives and evaluation criteria. The future belongs to knowledge-intensive food-production systems. In fact, the second Green Revolution has already started, through the scattered efforts of some researchers and farmers. The problem is how to bring this non-paradigmatic research to the attention of international organizations and national bodies and give it adequate support. Old paradigms die hard and the first Green Revolution still monopolizes most of the available resources.

References

Sachs, I. 1980. *Stratégies de l'écodéveloppement*. Paris: Editions Economie et humanisme/Editions ouvrières.

Sachs, I., A. Bergeret, M. Schiray, S. Sigal, D. Thery, and K. Vinaver, 1981. *Initiation à l'écodéveloppement*. Toulouse: Editions Privat. (Collection Regard). This collective book is the outcome of research carried out at CIRED, with the support of the United Nations Environment Programme (UNEP). It was prepared, under the direction of Ignacy Sachs, within the context of the Dag Hammarskjöld Project on Development and International Cooperation.

Sachs, I. 1984. The food–energy problem. In *Energy and agriculture: Interaction futures – policy implications of global models*, M. Levy and J. L. Robinson (eds), 25–40. New York and London: Harwood Academic Publishers for the United Nations University.

Contributors

Kurt Egger is professor at the Department of Botany at Heidelberg University. He is working in the field of ecologically oriented agrosystems and human ecological problems, with consultative activities in rural development projects.

Bernhard Glaeser is a project director at the Environmental Policy Research Unit (previously International Institute for Environment and Society) of the Wissenschaftszentrum Berlin für Sozialforschung (WZB). His research activities focus on environmental problems in developing countries and the theory of human ecology.

Edmund K. Oasa is a critical analyst of international agricultural research policy. He has held research positions at the East–West Center in Honolulu, Hawaii, the International Institute for Environment and Society of the Science Center, Berlin, and the Department of Sociology at the University of Kentucky.

Brigitta Martens worked in the field of integrated pest management until 1983, when she joined Professor Egger's research group. She is now a special consultant in community landscape planning.

Kevin Phillips-Howard, formerly visiting research fellow of the International Institute for Environment and Society, Science Center Berlin (WZB), is Associate Professor of Geography at the University of Juba, the Sudan.

Ademar Ribeiro Romeiro is fellow of the National Council of Scientific and Technological Development at the International Research Centre on Environment and Development in Brazil, on leave from the Brazilian Institute of Geography and Statistics.

Rathindra Nath Roy heads the Catalyst Group, a networking consulting firm in Madras, India, focusing on development and environmental analysis, planning, management and training.

Ignacy Sachs is Director of the Centre International de Recherche sur l'Environnement et le Développement, Ecole des Hautes Etudes en Sciences Sociales, Paris, and Programme Director, the United Nations University Food Energy Nexus sub-programme.

Pierre Spitz was with the United Nations Research Institute for Social Development (UNRISD) before he joined the Independent Commission on International Humanitarian Issues, Geneva. He is on leave from the National Institute of Agricultural Research (INRA) in Paris, where he is Research Director.

Index